A new fast and accurate grid deformation method for r-adaptivity in the context of high performance computing

Dissertation
zur Erlangung des Grades eines
Doktors der Naturwissenschaften

Dem Fachbereich Mathematik der Universität Dortmund
vorgelegt am 07. Dezember 2007 von

Diplom-Mathematiker Matthias Grajewski

Bibliografische Information der Deutschen Nationalbibliothek

Die Deutsche Nationalbibliothek verzeichnet diese Publikation in der
Deutschen Nationalbibliografie; detaillierte bibliografische Daten sind
im Internet über http://dnb.d-nb.de abrufbar.

ISBN 978-3-8325-1903-2

Logos Verlag Berlin GmbH
Comeniushof, Gubener Str. 47,
10243 Berlin
Tel.: +49 030 42 85 10 90
Fax: +49 030 42 85 10 92
INTERNET: http://www.logos-verlag.de

Contents

I

Chapter 1

Introduction

Although this thesis has only one author, which is common for a PhD thesis, the author hesitates to insist on using the first person in writing. Therefore, we decided to switch to the first person plural as *plurale humilitatis*. Where the presented results are joint work, i.e. where "we" indicates a group of authors in fact, it is explicitly stated.

First and foremost I would like to thank Prof. Dr. S. Turek for his supervision, the interesting topic and the possibility to just work in his group. Almost all the calculations in this thesis were performed using the FEM package FEAST, to which the author had the honour to participate. Thus, my special thank goes to the FEAST team with its inner circle consisting of Christian Becker, Sven Buijssen, Dominik Göddeke and Hilmar Wobker (and the author, of course). Last but not least I want to express my thank to Michael Köster for the many fruitful discussions about grid deformation methods.

1.1 A posteriori error control - state of the art

Numerical simulation turns out to be (or at least to become) a key tool in research and development in both the engineering and academic areas, as it enables insight into technical and physical processes which can be investigated experimentally only with huge effort (e.g. airplanes in flight) or cannot be investigated at all (e.g. dynamics of stellar forming). On the other hand, numerical simulation saves time and money in the development process of technical devices, as it allows to replace experiments by computations (see e.g. [53]). Based on an idea of Courant [33], after six decades of development the Finite Element Method (FEM) can be regarded as one of the most prominent methods for numerical simulation. However, like all other techniques for numerical simulation, the FEM provides approximate results only. Therefore, it is mandatory to know or at least to estimate the difference of the true, but unknown solution u and the computed approximate solution u_h in order to be able to assess the quality of the numerical simulation. Here, the emphasis is put on *a posteriori* error estimators η which incorporate u_h, $\eta = \eta(u_h) \geq 0$. Therefore, these estimators have to be applied

1

after the computation has been accomplished. Here and in what follows, we refer to computations on a bounded two-dimensional domain Ω with boundary $\partial\Omega$; if not, it is stated explicitly. The computational mesh $\mathcal{T}$ consists of $|\mathcal{T}| = NEL$ quadrilateral elements $T \in \mathcal{T}$ of size h_T. The global mesh width h is defined by $h := \max_{T \in \mathcal{T}} h_T$. Furthermore, we introduce the abbreviations $\mathcal{V}$ with $|\mathcal{V}| = N$ for the set of vertices and $\mathcal{E}$ for the set of edges in the grid.

In contrast to a priori error estimators, a posteriori error estimators obtain quantitative error bounds for a given FEM simulation and thus deserve our particular interest. In this thesis, error estimators are meant to be a posteriori error estimators; if this is not the case, it will be explicitly stated. An error estimator as numerical algorithm has to meet two substantial requirements:

- **reliability:** With $e_h := u - u_h$, the error $|J(e_h)|$ may be underestimated at most by a constant c independent of the element size:

$$\exists c \in \mathbb{R}, c > 0 : \quad \eta(u_h) \geq c\,|J(e_h)|. \tag{1.1}$$

- **efficiency:** The error may be overestimated at most by a constant C not depending on the element size:

$$\exists C \in \mathbb{R}, C > 0 : \quad \eta(u_h) \leq C\,|J(e_h)|. \tag{1.2}$$

Here, J denotes a continuous but not necessarily linear functional. To measure the quality of estimation, the *efficiency index*

$$I_{\text{eff}} := \frac{\eta(u_h)}{|J(e_h)|}$$

is computed. Trivially, $c \leq I_{\text{eff}} \leq C$ holds. If $I_{\text{eff}} \to 1$ for $h \to 0$, the estimator is called *asymptotically exact*. For practical reasons, it is desired that the error estimator can be written in the form $\eta = \sum_{T \in \mathcal{T}} \eta_T$ or $\eta = \left(\sum_{T \in \mathcal{T}} \eta_T^2\right)^{1/2}$ where η_T denotes the contribution of the cell T to the overall error. By this, one gains insight into the error distribution in space and thus the estimator can be utilised for controlling the grid adaptation process described in the next section.

After roughly thirty years of research on error estimators and error bounds, a plethora of different methods for a posteriori error control is available. In the following, we give a short overview on techniques for error control in the context of FEM. Here and in what follows, the Sobolev spaces $H^k := H^k(\Omega)$ and $H_0^k := H_0^k(\Omega), k \in \mathbb{N}$, are defined in the standard way [20] as well as the Lebesgue spaces $L^k := L^k(\Omega)$. For the definition of Sobolev spaces of fractional order $H^{k+\alpha} := H^{k+\alpha}(\Omega), 0 < \alpha < 1$, we refer to Babuška [3].

For the sake of simplicity, we consider the Poisson problem with homogenous Dirichlet boundary conditions

$$-\Delta u = f, \quad u|_{\partial\Omega} = 0, \tag{1.3}$$

and

$$(\nabla u, \nabla \varphi) = (f, \varphi) \quad \forall \varphi \in H_0^1 \tag{1.4}$$

in weak formulation, respectively. It serves as prototype for any scalar-valued linear elliptic boundary value problem $a(\cdot, \cdot) = l(\cdot)$. The term $(\cdot, \cdot)_D$ symbolises the standard L^2-scalar product on a domain $D \subseteq \Omega$. For $D = \Omega$, the subscript is omitted.

Most of the error estimators target the gradient error $(J(\cdot) := ||\cdot||_1)$ or the energy error $(J(\cdot) = (a(\cdot, \cdot))^{1/2})$, where $||\cdot||_1$ denotes the standard H^1-seminorm

$$||\cdot||_1 := ||\cdot||_{1,\Omega} := ||\nabla \cdot||_0 := ||\nabla \cdot||_{0,\Omega} := (\nabla \cdot, \nabla \cdot)^{1/2}.$$

Note that for the Poisson equation, energy and gradient error coincide. Starting with the pioneering work of Babuška and Rheinbold [5], many energy error estimators have been developed [1, 28, 32]. The energy error estimators can be roughly subdivided into two classes. The first class consists of the classical residual based estimators ([5, 9]) and the equilibrated residual estimators [1]. The former ones only use information about the local residual on a single element. Therefore, they are easy to implement in FEM codes and require only small computational effort. In contrast to this, the equilibrated residual estimators employ local auxiliary problems, which are small and hence can be computed fast in comparison to the numerical simulation. Albeit known as reliable and efficient (see the references cited above), the (unknown) constants c and C may be far away from one in the case of residual based error estimators. This prevents reliable error control in a practical sense, where the error is to fall below a prescribed tolerance TOL,

$$|J(e_h)| \leq TOL.$$

However, for many practical computations – at least in the case of linear problems – both constants are rather close to one. Therefore and because of their easy implementability, this kind of estimators is very popular. Besides the Poisson equation, residual based error estimators have been developed for convection-diffusion equations (e.g. [78]), the incompressible Stokes equation [29, 34], for problems in linear elasticity [30] and elastoplasticity [26], for nonlinear equations [77], instationary problems (e.g. [79]) as well as for numerous other applications. Various authors investigated anisotropic variants of residual based estimators [30, 34, 48, 55]. Moreover, they are available for non-conforming and mixed FE as well (see e.g. [28] and the references cited therein). Residual based error estimators are at hand for two-and three dimensional problems.

The second class of energy error estimators contains besides many other ones the widespread ZZ-technique (also referred to as SPR-method) [85, 86, 87, 88] and the newer PPR-method [84]. This sort of error estimators is denominated as reconstruction-based. They aim to replace the unknown term ∇u in the gradient error $(\nabla(u - u_h), \nabla(u - u_h))^{1/2}$ by a *recovered gradient* $G_h(u_h)$. The recovered gradient is computed applying certain interpolation- or projection operators to u_h.

As an example of a reconstruction-based method, let us consider bilinear conforming Finite Elements (Q_1) and denote for a given vertex v the set of all elements adjacent to v by $\mathcal{T}(v)$. For the ZZ-technique, $\nabla(u_h)$ is evaluated in the centers of the elements in $\mathcal{T}(v)$. The recovered gradient G_v in v is defined as bilinear interpolation of these values. The corresponding linear system of equations is solved in a least squares sense, if $|\mathcal{T}(v)| \neq 4$. The recovered gradient for an arbitrary point is defined as the unique function in $Q_1(\mathcal{T})$ which fulfils $G_h(u_h)(v) = G_v \; \forall v \in \mathcal{V}$ where $\mathcal{V}$ symbolises the set of vertices in the grid $\mathcal{T}$. Here and in the following, the H^1-conforming Finite Element space consisting of piecewise polynomials of order k in each argument on a grid $\mathcal{T}$ is abbreviated by $Q_k(\mathcal{T})$.

The PPR-method instead evaluates u_h in all vertices in $\mathcal{T}(v)$ in order to compute a biquadratic interpolation polynomial P on this patch. If $|\mathcal{T}(v)| \neq 4$, the corresponding linear system is solved in a least-squares sense. The recovered gradient in v is set to $\nabla P(v)$, the recovered gradient $G_h(u_h)$ is defined like in the case of the ZZ-technique.

For some partial differential equations (PDEs) in two dimensions, this recovered gradient features *superconvergence* under rather restrictive conditions on the smoothness of the solution and the mesh, i.e. the error of $G_h(u_h)$ decreases at a higher rate than the gradient itself (see the references above). Superconvergence effects in three dimensions are very rare [21, 56]. If the recovered gradient shows superconvergence, the error estimation is asymptotically exact. However, for most of the problems considered in this thesis the requirements for superconvergence are not granted. Nevertheless, under mild conditions all reconstruction based error estimators remain efficient and reliable [10, 25, 27]. This result remains valid in three dimensions.

Many numerical simulations, however, are carried out to approximate the values of *derived quantities* such as lift or drag values in fluid dynamics. In this situation, one is consequently not interested in estimating the error in some global norm, but in estimating the error of the derived quantity itself. In the framework of equations (1.1) and (1.2), J in this situation represents an *evaluation functional* defining the computation of the derived quantity from the solution. Popular examples of evaluation functionals are

- $J_{x_0}(\varphi) := \varphi(x_0)$ (point evaluation),

- $J_\Gamma(\varphi) := \int_\Gamma \partial_n \varphi ds$ (evaluation along a line),

- $J_S(\sigma) = \int_{\Gamma_D} \mathfrak{n}^\top \sigma \mathfrak{n} \, ds$ (mean stress value along a line),

- $J_{D/L}(\varphi, \chi) := \int_\Gamma \mathfrak{n} \cdot \sigma(\varphi, \chi) \cdot \mathfrak{e}_{x/y} \, ds$ (lift /drag computation).

Here, $\mathfrak{n}$ stands for the unit vector, whereas $\mathfrak{e}_{x/y}$ symbolises the unit vector in x- and y-direction, respectively.

Although in the case of continuous functionals J it is possible to derive upper and lower bounds of $|J(e)|$ in terms of the energy error, these bounds rely on unknown constants which are usually hard to determine and which can be very large. Because of this, the aforementioned energy error estimators are not suitable to estimate $|J(e_h)|$ in practical computations.

These aforementioned difficulties in estimating derived quantities can be overcome by introducing the *dual problem* and using its (unknown) solution as ingredient for an upper bound for the error. Johnson and his coworkers obtain such an estimate by applying generic estimates to the norm of the dual solution [38] and thus avoid computing the dual problem. In contrast to this, Becker and Rannacher [7, 66] compute the dual problem approximately and utilise the approximate dual solution in their estimators. For a linear and H^1-continuous functional, they proceed for the Poisson equation with zero Dirichlet boundary conditions as follows to obtain sharp and computable bounds for $|J(e)|$. Defining the *dual problem*

$$(\nabla z, \nabla \varphi) = J(\varphi) \quad \forall \varphi \in H_0^1, \tag{1.5}$$

one obtains

$$
\begin{aligned}
|J(u - u_h)| &= |(\nabla u, \nabla z) - (\nabla u_h, \nabla z)| \tag{1.6}\\
&= |(f, z - \varphi_h) - (\nabla u_h, \nabla(z - \varphi_h))| \quad \forall \varphi_h \in V_h \\
&= \left| \sum_{T \in \mathbf{T}} (f, z - \varphi_h)_T - (\nabla u_h, \nabla(z - \varphi_h))_T \right| \forall \varphi_h \in V_h \\
&\approx \left| \sum_{T \in \mathbf{T}} (f, \tilde{z} - z_h)_T - (\nabla u_h, \nabla(\tilde{z} - z_h))_T \right|. \tag{1.7}
\end{aligned}
$$

The expression $\tilde{z}$ stands for a suitable approximation to the unknown *dual solution* z, z_h symbolises an approximation to the dual solution in the FEM ansatz space. Applying Green's formula to (1.7), the error in the target functional is represented by

$$
\begin{aligned}
|J(u - u_h)| &\approx \left| \sum_{T \in \mathbf{T}} (f + \Delta u_h, \tilde{z} - z_h)_T - \frac{1}{2}([\partial_\mathbf{n} u_h], \tilde{z} - z_h)_{\partial T \backslash \partial \Omega} \right| \\
&\leq \left| \sum_{T \in \mathbf{T}} \|f + \Delta u_h\|_T \, \|\tilde{z} - z_h\|_T \right. \\
&\qquad \left. + \frac{1}{2} \|[\partial_\mathbf{n} u_h]\|_{\partial T \backslash \partial \Omega} \|\tilde{z} - z_h\|_{\partial T \backslash \partial \Omega} \right| \tag{1.8}
\end{aligned}
$$

Here, for two elements T_1 and T_2 sharing the edge T with normal vector $\mathbf{n}$, $[\partial_\mathbf{n} u_h] := \partial_\mathbf{n} u_h|_{T_1} - \partial_\mathbf{n} u_h|_{T_2}$ denotes the jump of the normal derivative of u_h on E. As in (1.8) the local residuals are weighted with the contributions of the dual solution, this method is called *dual weighted residual based method*, DWR-method [7, 66] in short. In the literature, different approaches are proposed to obtain $\tilde{z}$.

1. **FEM-approaches:** The function $\tilde{z}$ is obtained as solution of an FEM discretisation of (1.5). This can be done by using the same grid as the actual

problem ("primal problem") but by employing higher order elements. Becker and Rannacher [66] compute the primal problem using bilinear conforming elements and the dual problem employing conforming biquadratic elements. Doing so, their numerical experiments show that an asymptotically exact error estimation is achieved. On the other hand, the computational amount for error estimation dominates the one for the primal problem. As an alternative, the dual solution can be computed using the same Finite Element like the primal one, but on a further refined mesh which again leads to demanding computations for error estimation only.

2. **Reconstruction-based approaches:** At first, z_h is computed in the same ansatz space like u_h. After this, one applies reconstruction and interpolation techniques to z_h to gain $\tilde{z}$. This way, one ends up with an error estimator which can be computed with reasonable numerical effort. Becker and Rannacher suggest a biquadratic interpolation of z_h on a patch consisting of four elements [66]. In contrast to this, Giles and Pierce apply spline-interpolation to construct $\tilde{z}$ [44, 45]. Another variant of the reconstruction approach is proposed by Neittaanmäki, Korotov and Martikainen [61]. Considering the Poisson equation, they replace the terms ∇u and ∇z by $G_h(u_h)$ and $G_h(z_h)$ in formula (1.6) via applying the ZZ-technique. Their experiments show that this estimator exhibits asymptotic exactness for certain test cases. A similar technique is applied by Ovall, who proves under very (and in practical situations even unrealistically) strict conditions on the smoothness of u and z the asymptotic exactness of his proposed goal-oriented error estimator [62].

In the area of DWR, rapid progress has been made in the last years, and this method is no way restricted to the Poisson equation. Extensions to nonlinear problems like the Navier-Stokes equations have been developed by Becker and Rannacher [65] as well as the application to nonlinear evaluation functionals. Variational inequalities in the field of elastoplasticity and obstacle problems are treated by Blum and Suttmeier [16, 17]. Hoffman and Johnson apply the DWR-method to the instationary Navier-Stokes-equation and thus extend the technique to time-dependent problems [51, 52]. Heuveline and Rannacher develop an DWR method for hp-adaptive FEM discretising the Poisson equation [50]. Formaggia, Micheletti and Perotto propose an anisotropic variant of the DWR-method [40, 41]. These estimators provide information not only regarding the cell size but regarding cell shape and position for the adaptation process, too.

For the DWR method, there is no rigorous proof of reliability and efficiency the author is aware of. This is related to the arbitrary replacement of z by $\tilde{z}$. How this replacement affects the accuracy of the estimation is investigated by Suttmeier for the Poisson equation [70, 71, 72]. The insertion of $\tilde{z}$ in formula (1.6) leads to the equation

$$J(e) = a(u - u_h, z - \tilde{z}) + (f, \tilde{z} - z_h) - a(u_h, \tilde{z} - z_h).$$

Neglecting the term $a(u - u_h, z - \tilde{z})$ yields the estimator (1.7). Suttmeier computes $\tilde{z}$ as FEM solution of the dual problem using conforming bilinear elements on a finer mesh [72, 70, 71]. The term $a(u - u_h, z - \tilde{z})$ is controlled by gradient error estimation:

$$a(u - u_h, z - \tilde{z}) \leq ||u - u_h||_1 \, ||z - \tilde{z}||_1 \leq \eta_p \eta_d.$$

Obviously, Suttmeier's error estimator is reliable, if the gradient error estimators are reliable.

If the error of the evaluation function has been estimated well enough, this information can be used to perform *a posteriori error correction*. If J is a linear functional, one can define a better approximation to $J(u)$ by

$$J(u) = J(u - u_h + u_h) = J(e_h) + J(u_h) \approx \eta(u_h) + J(u_h) =: J^{\text{corr}}(u_h).$$

Numerical experiments by Giles and Pierce [44, 45] and Suttmeier (see above) show, that the error of the corrected value $J^{\text{corr}}(u_h)$ under certain conditions may decrease even with a higher order than the error of $J(u_h)$ itself. However, there is no computable sharp bound for the error of $J^{\text{corr}}(u_h)$ yet. Extensions to nonlinear functionals via Taylor expansions have been developed by Giles and Pierce [44].

1.2 Grid modification techniques in adaptive FEM

If after a numerical simulation the a posteriori error estimator shows that the prescribed error tolerance has not been achieved, the computation has to be repeated using a different FEM space. This can be constructed by refining the mesh (*h-adaptivity*), by relocating the grid points (*r-adaptivity*) or by adjusting the polynomial degree of the basis functions on the same mesh (*p-adaptivity*). Every grid adaptation algorithm aims at equilibrating the error distribution on the grid. For a detailed derivation of the equilibration principle, we refer to [7].

In the following subsections, we will give a short overview of these three techniques. The last of the sections deals with combinations of the aforementioned methods.

1.2.1 *h*-adaptivity

Using the information about the error distribution which is provided by the error estimator, the grid is locally refined in the regions of large error contributions and locally coarsened in the region of small error contributions. Assuming a quadrilateral grid, the grid is locally refined by subdividing the elements in that region into four new elements each one or several times. Usually, this is done isotropically by connecting the midpoints of opponent element edges. In this thesis, *h*-adaptivity is performed this way only. Doing so, all coordinates of the vertices in the mesh remain fixed. By the local subdivision, "hanging nodes"

8

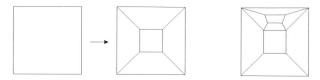

Figure 1.2.1: Buffer elements for h-refinement, degeneration of elements during iterated naive refinement

occur on the boundaries of regions with different level of refinement, i.e. nodes with no counterpart on the coarser edge-adjacent element. In the vast majority of FEM codes, the refinement and coarsening is performed element-wise. For practical reasons, the difference of the levels of refinement of two edge-adjacent cells is usually bounded to one. In this case, there is at most one hanging node per element edge.

Local refinement on quadrilateral meshes does not require hanging nodes. They can be avoided employing buffer elements (compare figure 1.2.1, left). Unfortunately, the naive insertion of buffer elements leads to degenerated elements after several refinement steps (figure 1.2.1, right). In the past, more sophisticated strategies have been developed to avoid degenerated elements due to local refinement, but these strategies are usually complicated and tedious to implement (compare e.g. [67]). In three dimensions, it is very hard to perform h-adaptivity with buffer elements, as in a suitable grid refinement algorithm, more than hundred different cases requiring different buffer elements have to be considered [68]. Therefore, this kind of h-refinement will not be considered in this thesis, and h-adaptivity will be implicitly connected to hanging nodes in what follows. It should be noted that on triangular grids, hanging nodes can be circumvented in local refinement by applying sophisticated strategies involving buffer elements, e.g. the "red-green-blue-refinement" by Bank, Sherman and Weiser [8].

Bangerth and Rannacher propose three different strategies for h-adaptivity [7]. The authors assume that the error estimator can be written as the sum of element-wise error contributions η_T,

$$\eta = \sum_{T \in \mathcal{T}} \eta_T \, ,$$

and the usage of hanging nodes on quadrilateral meshes.

- **fixed fraction:** The element-wise error contributions η_T are sorted according to their size. Then, one chooses two constants $0 \le a, b \le 1$ which fulfil $1 - a > b$. After this, the indices N_1 and N_2 are computed, such that

$$\sum_{i=1}^{N_1} \eta_{T_i} = a\eta, \quad \sum_{i=N_2}^{NEL} \eta_{T_i} = b\eta \qquad (1.9)$$

holds. The elements with indices from one to N_1 are refined, the elements with indices from N_2 to *NEL* are coarsened. The shortcoming of this strategy is its dependency of the parameters a and b, which have to be chosen heuristically.

- **error-balancing:** This strategy aims to fulfil the relation

$$\eta_{T_i} \approx \frac{TOL}{N_e},$$

where N_e denotes the number of elements on the adapted grid. As N_e is not known a priori, error-balancing is an implicit strategy which thus requires more effort than the aforementioned fixed fraction strategy. Therefore, this strategy is applied in rare cases only and will not be considered in this thesis.

- **mesh optimisation:** This strategy relies on analytical considerations in contrast to the aforementioned strategies. Starting from the assumption that a continuous mesh size function $h(x)$ and a continuous error functional $\eta(h)$ exist, one solves the optimisation problem

$$\eta(h) \to \min, \quad N(h) \leq N_{\max}.$$

In its solution $h_{\mathrm{opt}}(x)$ all information regarding grid refinement is encoded. In contrast to the approaches described previously, one gains information about the necessary amount of refinement, such that under certain conditions an optimal grid can be created in only one iteration step. However, for practical calculations, these conditions are often violated.

Even though h-adaptivity is widely used in the FEM community for grid adaptation, there is surprisingly little knowledge about the convergence of error-driven adaptive FEM (AFEM). The first construction of an AFEM with a convergence proof for the Poisson equation discretised by linear conforming Finite Elements was provided by Dörfler [37]; Morin, Nochetto and Siebert proved convergence for the same problem under less restrictive assumptions [60]. Very recently, Carstensen and Hoppe showed the same result for the Poisson equation discretised by the nonconforming Crouzeix-Raviart element [31]. All articles dealing with convergence of AFEM, at least the ones we are aware of, are restricted to triangle meshes without hanging nodes. Because of this, for the class of meshes considered in this thesis, no convergence proof of AFEM exists at all. Moreover, there is no convergence proof for any AFEM with respect to the error of derived quantities.

1.2.2 *r*-adaptivity

In contrast to h-adaptivity described in the preceding subsection, no new grid nodes are introduced when applying r-adaptivity. Instead, the existing grid nodes are relocated while preserving the topology of the mesh. The redistribution of the grid nodes is formalised by applying a deformation mapping $\Phi(x, y) = (\xi, \eta)^\top$ to the coordinates of the grid points. Here, (x, y) denote the *computational coordinates*, i.e. the coordinates on the initial domain or on an auxiliary domain from which the *physical coordinates* (ξ, η) are computed or created. The computation of the actual numerical simulation takes place in physical coordinates. Variational approaches (see e.g. [23]) obtain Φ by minimising a functional I depending on Φ. Many approaches of choosing I have been investigated in the literature which are usually motivated by geometric considerations. Cao, Huang and Russell consider the fairly general functional

$$I(x, y) = \frac{1}{2} \int_\Omega \frac{1}{\sqrt{m}} \left(\nabla x^\top G^{-1} \nabla x + \nabla y^\top G^{-1} \nabla y \right) d\xi \, d\eta, \qquad (1.10)$$

which contains many of the choices for I in earlier literature as special cases and thus provides a natural framework for comparing various earlier methods of mesh deformation. The positive definite symmetric 2×2-matrix G is named *monitor function*. The stationary point of I usually is characterised by the solution of a nonlinear PDE emerging from applying the *Euler-Lagrange equations*

$$\operatorname{div}(G^{-1} \nabla x) = 0 = \operatorname{div}(G^{-1} \nabla y), \qquad (1.11)$$

to I. This nonlinear PDE is then attempted to be solved numerically in practical computations. Note that the nonlinearity of the resulting PDEs emerges from reformulating the linear Euler-Lagrange equations (1.11) in the desired coordinate set (ξ, η), where the interplay of the physical coordinates (ξ, η) and the computational coordinates (x, y) is implicitly determined by the functional I.

Winslow's method [81], which is one of the earliest grid deformation techniques we are aware of, appears as special case of (1.10) with the setting $G = w(\xi, \eta)Id$. Here, Id denotes the unit matrix. The corresponding functional is

$$I(x, y) = \frac{1}{2} \int_\Omega \frac{1}{w(\xi, \eta)} \left(|\nabla x|^2 + |\nabla y|^2 \right) d\xi \, d\eta.$$

The function w plays the role of a weighting function and has to be prescribed by the user. "The direction where w changes most rapidly is the one in which mesh lines are compressed or expanded most" [23]. Historically, Winslow interprets the deformation problem "as a potential problem with the mesh lines playing the role of equipotentials" [81] and defines the mesh nodes as the intersections of these lines. As in their computation a Laplace problem in (x, y) is involved and "because of the well-known averaging property of solutions to Laplace's equation, we might expect a mesh constructed in this way to be, in some sense, smooth" [81]. Thus, Winslow's choice of I is motivated by optimising the smoothness of the resulting grid. Taking

$$G(x, y) = \frac{M(x, y)}{\sqrt{\det(M(x, y))}},$$

M positive definite and symmetric, one ends up with the *method of harmonic mappings* [19] by Brackbill and Saltzman with the functional

$$I(x,y) = \frac{1}{2} \int_\Omega \frac{1}{\sqrt{\det(M(\xi,\eta))}} \left(\nabla x^\top M^{-1} \nabla x + \nabla y^\top M^{-1} \nabla y \right) d\xi \, d\eta.$$

This method was originally developed as extension of Winslow's method, where not only the smoothness of the resulting mesh is to be maximised but additionally the volume-weighted orthogonality and the weighted volume variation are taken into account. The resulting functional is constructed as a weighted sum of these geometric quantities. The majority of the variational grid deformation methods leads to nonlinear partial differential equations which are often very hard to solve. Moreover, in many cases "it is usually very hard to predict the overall resulting behaviour from the functional itself" [23] and thus it is not clear how to choose the monitor function in order to control the grid adaptation process. It is not even clear which functional I to choose from the plenty proposed in literature. Because of these disadvantages, we decided not to study this kind of deformation methods in detail.

A different, less complicated approach is developed by Fleitas, Jiang and Liao [39]. Considering a time-dependent scalar monitor function

$$f(x,y,t) > 0 \quad \forall \, (x,y) \in \bar{\Omega}(t), \, t \in [0,T]$$

with

$$\int_{\Omega(t)} \frac{1}{f(x,y,t)} dx = |\Omega(0)|,$$

the time-dependent mapping $\Phi(x,y,t)$ is defined by

$$|J\Phi(x,y,t)| = f(\Phi(x,y,t),t) \quad \forall \, 0 \le t \le T,$$

where $J\Phi$ denotes the Jacobian matrix of Φ and $|\cdot|$ its determinant. The aim of this approach is control the *distribution of the element size* on the resulting grid which is governed by $|J\Phi|$. In contrast to the aforementioned approaches which lead to the demanding minimisation of complicated functionals, here the mapping Φ is constructed by solving a global Poisson problem and several decoupled ordinary differential equations (ODEs). As our Finite Element package FEAST provides sophisticated and fast Poisson solvers, Liao's method appears as natural starting point for our own investigations in the field of r-adaptivity and grid deformation.

1.2.3 *p*-adaptivity

Let us assume that the Finite Element chosen is member of a family of Finite Elements with increasing numbers of unknowns per geometrical element. A common example of this is the family $\mathcal{Q}_k$ where the ansatz space consists of polynomials of order k. In p-adaptive algorithms, the grid remains fixed during the adaptation process. The enrichment of the FEM space is provided by taking the "next higher" member in the family of Finite Elements. It has been shown [6] that this way extremely accurate approximations to the solution can be achieved. However, there are some drawbacks. Increasing the polynomial degree (p-refinement) leads to matrices which are far less sparse than the ones obtained by refining the grid (h-refinement) and which possibly take substantial amount of memory to store. Moreover, the development of fast iterative solvers for the algebraic systems arising, in particular the development of suitable preconditioners, is a challenging task (compare e.g. [13] and the references cited therein). Using a few elements only, it is very hard or even impossible to reproduce geometric features of complex domains. It is a challenge to resolve local features of the solution by locally refining the FEM space, as enriching the polynomial space affects a possibly large part of the domain. Therefore, it is very difficult to establish sophisticated adaptivity concepts in the context of p-FEM. Because of this, the strength of p-adaptivity can not be exploited in practice and thus p-adaptivity in the pure form described here is rarely used in FEM tools. In this thesis, this method will not be considered in more detail.

1.2.4 Combined approaches to grid adaptivity

The most common combination of adaptivity methods in practical FEM codes is the combination of h- and p-adaptivity, usually denoted by hp-adaptivity. The systematic investigation of this FEM variant started in 1981 by Babuška and Dorr [4] and revealed possible *exponential convergence* on suitable grids and for appropriate choices of the polynomial degree [6]. Exponential convergence cannot be achieved by any h-adaptive FEM. Therefore, the theoretical potential of the hp-method exceeds by far that of the pure h-method. Nevertheless, exponential convergence requires an extremely smooth solution and detailed a priori knowledge about the solution in order to construct the grid and the distribution of the polynomial degree. Demkowicz and his coworkers [36] developed adaptive refinement strategies for both h- and p-refinement based on a posteriori error estimates and apply their hp-adaptive method successfully to scalar elliptic problems as well as the Maxwell's equations. Schröder and Blum develop and apply an hp-adaptive algorithm with a posteriori error control to contact problems occurring in belt grinding processes [69]. Heuveline and Rannacher develop an hp-adaptive method for estimating derived quantities of the solution of the Poisson equation [50]. However, despite these efforts, the theory especially regarding error estimators for hp-methods seems to be in a rather immature state yet compared to the knowledge available for the h-method, and many practical problems remain to be resolved, e.g. the development of fast robust solvers. Moreover, the implementation is very involved and leads to practical challenges. Establishing an error-driven hardware-oriented hp-algorithm working on parallel machines without significant speed losses would lead to a overwhelming number of questions and open problems and is therefore well beyond the scope of this thesis.

NEQ	SP-MOD	SP-STO	SBB-V
4,225	557	561	1805
66,049	395	223	660
1,050,625	391	75	591

Table 1.3.1: Measured MFLOP/s -rates for sparse matrix-vector multiplication in the case of different matrix structures, computed with an Opteron 852; taken from Becker [12].

1.3 High performance computing and FEAST

Many commercial and scientific software packages for FEM simulation make use of the error estimation procedures introduced above and combine them with sophisticated strategies of grid adaptation e.g. the ones described in the preceding chapters. This is done in order to reduce the number of unknowns which are necessary to reach the prescribed accuracy in the numerical simulation as far as possible. In many cases this strategy leads to impressive improvements compared to unadopted computations. Obviously, the *discretisation efficiency*, i.e. the ratio of number of unknowns and accuracy, can be vastly improved by adaptive methods. However, for practical computations the number of unknowns is less important than the computational time which is of course related to the number of unknowns to some extent. Therefore, not the number of equations alone but the computational effort in terms of arithmetic operations and data accesses required for the solution has to be minimised. This justifies the research for numerically efficient numerical components, mainly the development of fast solvers. Last but not least it is mandatory that the tremendous computing power today's computers offer can be fully exploited for the desired numerical simulation. Thus, we define as *overall efficiency* the ratio of computational time and accuracy. It is obvious that an FEM-code suitable for challenging problems has to offer overall efficiency: If the discretisation lacks efficiency, the accuracy requirements lead to overwhelmingly large numbers of equations; the best discretisation is useless, if the arising systems of equations cannot be solved in reasonable time; if the computer shows extremely weak performance in the simulation, the computation lasts much longer than anticipated. Surprisingly, almost no current simulation tool offers all three types of efficiency altogether. Many FEM codes coming from mathematical research feature extremely efficient discretisations, as this aspect is one of the main topics in FEM theory, but often lack overall efficiency, as the focus is not put on advanced and sophisticated solvers in these packages. In particular, hardware efficiency is usually neglected at all. On the other hand, many codes coming from engineering groups exhibit the ability of exploiting the strength of high performance computers, i.e. offer high hardware efficiency, but often the discretisation efficiency is poor, as no reliable and efficient error control and sophisticated grid adaptation mechanisms are available.

In his PhD thesis [12], Becker examines the hardware efficiency of modern FEM software. As representative FEM code, he considers the FEM package FEAT [14] which is regarded to be a mature and efficient FEM library. It serves as basis for the CFD-solver FEATFLOW [73]. As prototypical operation in FEM, he analyses the multiplication of a sparse matrix stored in the standard CSR-format and a vector. The matrix comes from discretising the Poisson problem on the unit square with conforming bilinear elements on the given grid. In table 1.3.1, the number of computing operations per second experienced in reality are displayed for different problem sizes and two kinds of grid numbering schemes. The number of equations is denoted by *NEQ*. Here, SP-MOD represents a regularly refined, but initially unstructured grid, and SP-STO stands for an FEM matrix in an elementwise h-adaptive code. The CPU used in this test, an AMD Opteron 852, has the theoretical capability of 4280 MFLOP/s ("peak performance"). The findings by Becker (compare table 1.3.1) reveal that matrix-vector multiplication using "classical" sparse techniques runs by far slower than one could expect. In the best case, only 15% of the computational power theoretically at hand is used in reality. Moreover, the MFLOP/s-rates decrease during grid refinement. This decay is most prominent for SP-STO. Consequently, the lowest speed is observed for the matrix structure coming up in codes employing elementwise h-adaptivity (SP-STO, only ca. 2% of the peak performance for $NEQ = 1,050,625$). Note that elementwise h-adaptivity is by far the most common grid adaptation technique. Obviously, increasing the discretisation efficiency by elementwise h-adaptivity leads to decreasing the hardware efficiency. In the implementation on the matrix-vector multiplication denoted by SBB-V however, the band structure of the matrix is exploited. In conjunction with sophisticated blocking techniques, this implementation yields at least ca. 15% of the peak performance which is a big improvement compared to SP-MOD and SP-STO. In a series of papers (e.g. [2, 74, 76]), the reasons for that behaviour were examined and possible remedies discussed.

One of the reasons for the observed lack in hardware efficiency in current (h-adaptive) FEM is the usage of unstructured grids which require indirect addressing. Thereby, they prevent the CPU from using its cache effectively. This results in processors waiting for the data from the main memory during the computation. This result indicates further, that the "traditional" widespread technique of elementwise h-adaptivity is not well-suited to modern computer architectures and that *new concepts of adaptivity must be developed* if one wants to use the full power of modern processors.

On the other hand, Becker showed that for *logical tensor product meshes* more hardware-efficient matrix-vector multiplication is possible (SBBLAS-technique), if the matrix which consists for line-wise numbering of nine bands, is stored bandwise. However, the MFLOP/s-rates of 15% of the peak performance only (compare table 1.3.1) indicate that still the memory access limits the computational speed. The exploit of the logical tensor product structure leads to the grid concept of FEAST, which we will describe shortly now.

A grid in FEAST consists of *macros*, which are logical tensor product meshes. A

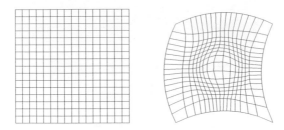

Figure 1.3.1: Tensor product mesh (left) and logical tensor product mesh (right)

grid is said to be a logical tensor product mesh, iff every inner node has exactly four neighbouring nodes (compare figure 1.3.1). For every such macro, the local matrix is built up exploiting the line-wise local numbering in each macro. This allows to use the SBBLAS-technique locally and enables high-speed matrix-vector multiplications. For details, we refer to [12]. As it is not possible to triangulate an arbitrary domain with a logical tensor product grid, the global grid consists of several macros. During the calculation, the number of macros remains fixed and does in particular not change during the refinement process of the grid. The triangulation on macro level is required to be conforming, i.e. the intersection of two macros has to be either empty or a vertex or a whole edge. In figure 1.3.2, a typical FEAST-grid is displayed. It becomes evident that the macros in this FEAST grid are logical tensor product grids only and are not restricted to a rectangular shape. As shown, even anisotropic refinement can be employed e.g. in order to better resolve boundary layers in CFD without destroying the local logical tensor product structure of the grid.

The SBBLAS-technique described above is one part of the FEAST-concept only. Apart of SBBLAS, FEAST is designed to be run on massively parallel distributed memory machines (clusters). In order to minimise the communication time during the solution process, which dominates the computational time in many other parallel codes, FEAST features the ScaRC-solver concept (Scalar Recursive Clustering). Designed as black box solver, ScaRC itself does not influence the design of FEAST-grids and therefore not the grid adaptation concepts in FEAST. Because of this we do not go into details here. Instead, we refer to Ch. Beckers's [12] and S. Kilian's [54] PhD theses, in which the ScaRC concept is developed. Becker shows the outstanding scalability properties of the FEAST solvers by excessive numerical testing. Expressed in the terminology introduced above, FEAST raises its hardware efficiency by exploiting the parallelism of modern super computers. Moreover, FEAST is capable to utilise specialised hardware as numerical coprocessors. Using commercially available graphics cards originally intended for computer gaming as coprocessors in FEAST, Göddeke experiences significant speed-ups of the solver times compared with computations on CPUs

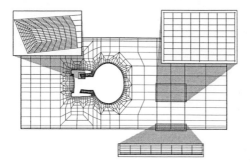

Figure 1.3.2: Prototypical FEAST grid

only. Using 160 graphics cards in parallel, Göddeke and his co-authors [46] discretise a Poisson problem with bilinear conforming elements on a tensor product mesh and solve the corresponding linear system with one billion unknowns in less than 50 seconds.

To maintain the superior computational speed FEAST is characterised by, it is mandatory as pointed out above that every concept of grid adaptation in FEAST *preserves the local logical tensor product structure of the macros.* Natural approaches for such an concept are

- **macro-wise h-adaptivity:** The grid is adjusted by the selected regular refinement of single macros introducing hanging nodes on macro edges. For practical reasons, the difference of the levels of refinement of two edge-adjacent macros is bound to one.

- **anisotropic refinement:** During the grid refinement, the elements on the coarse grid are subdivided anisotropically. Here, the refinement of edge-adjacent macros has to be adjusted in order to maintain the conformity of the grid.

- **grid deformation:** As pointed out before, grid deformation preserves the logical tensor product structure of a grid. Therefore, applying r-adaptivity by grid deformation to a grid consisting of several logical tensor product grids will lead to a grid consisting of several logical tensor product grids and thus to a valid FEAST-grid.

The development, analysis and realisation of an error-driven MFLOP/s-rate preserving grid adaptation concept in the FEAST context aiming at an FEM concept which allows for overall efficiency is the main goal of this thesis.

Chapter 2

A new method for grid deformation

In this section, we introduce our new deformation method which serves as a tool for r-adaptivity. It contains the method by Liao as special case. We mathematically prove that our method produces grids with the desired properties and give an overview over the mathematical context of our new deformation method.

Here and in the following the computational mesh is assumed to be conforming, i.e. no hanging nodes are allowed. The area of an element T is denoted by $m(T)$. The function space of k-fold continuously differentiable functions on the domain D is referenced by $\mathcal{C}^k(D)$ and $\mathcal{C}^k := \mathcal{C}^k(\Omega)$. For an interval I, $\mathcal{C}^{k,\alpha}(I)$, $0 < \alpha < 1$, denotes the space of functions with Hölder-continuous k-th derivatives. A domain is said to have an $\mathcal{C}^{k,\alpha}$-smooth boundary, iff the boundary can be parameterised by a function in $\mathcal{C}^{k,\alpha}(I)$. The Jacobian matrix of a smooth mapping $\Phi : \Omega \to \Omega$ is denoted by $J\Phi$, its determinant by $|J\Phi|$.

To formalize the deformation process, a weighting function $0 < \varepsilon < g \in \mathcal{C}^1(\bar{\Omega})$ and a *monitor function* $0 < \varepsilon < f \in \mathcal{C}^1(\bar{\Omega})$ are introduced. Both functions must be strictly positive in $\bar{\Omega}$. The reason why f is called monitor function will become clear below. Parts of this section have been previously published [47].

2.1 Derivation of the new grid deformation method

The theoretical background of the described approach – like Liao's approach [18, 22, 57, 58] – is based on Moser's work [35].

The aim of the numerical grid deformation algorithm described below is to construct a bijective transformation $\Phi : \Omega \to \Omega$ which satisfies

$$g(x)|J\Phi(x)| = f(\Phi(x)), \quad x \in \Omega \tag{2.1}$$

as well as

$$\Phi : \partial\Omega \to \partial\Omega. \tag{2.2}$$

If such a transformation Φ has been found, the new coordinates $X = (\xi, \eta)$ of a grid point x are computed by

$$X = \Phi(x).$$

Applying the area formula to an element T yields

$$m(\Phi(T)) := \int_{\Phi(T)} 1 \, dx = \int_T |J\Phi(x)| dx,$$

and employing the 1×1-Gauss rule in conjunction with formula (2.1), one obtains

$$g(x_c) \frac{m(\Phi(T))}{m(T)} = f(\Phi(x_c)) + \mathcal{O}(h).$$

Here, x_c stands for the center of T. Notice that the 1×1-Gauss rule is of second order for functions being in $\mathcal{C}^2$, but of first order only for $f \in \mathcal{C}^1$. If the function g represents the distribution of the element area in the mesh up to a spatially fixed constant $c(h)$, i.e.

$$g(x) = c(h) \, m(T) + \mathcal{O}(h), \; x \in T, \tag{2.3}$$

then one ends up with

$$c(h) \, m(\Phi(T)) = f(\Phi(x_c)) + \mathcal{O}(h), \tag{2.4}$$

if we assume that $m(T)$ and $m(\Phi(T))$ have the same order. This will be proven in lemma 3.1.7. Thus by prescribing the monitor function f, the element T will get – up to a spatially fixed scaling constant – the size defined by the value of f in the position of the image of T in the deformed grid.

Condition (2.2) requires that the domain itself does not change, i.e. $\Phi(\Omega) = \Omega$. Using formula (2.1) and the transformation rule, we have

$$\int_{\Phi(\Omega)} \frac{1}{f(x)} dx = \int_\Omega \frac{1}{f(\Phi(x))} |J\Phi(x)| \, dx = \int_\Omega \frac{1}{f(\Phi(x))} \cdot \frac{f(\Phi(x))}{g(x)} \, dx = \int_\Omega \frac{1}{g(x)} \, dx$$

which under our assumption $\Phi(\Omega) = \Omega$ yields the *compatibility condition*

$$\int_\Omega \frac{1}{f(x)} dx = \int_\Omega \frac{1}{g(x)} dx$$

to f and g and thus determines the scaling factor $c(h) = ch^{-2}$. Therefore, equation (2.4) can be rewritten as

$$m(\Phi(T)) = ch^2 f(\Phi(x_c)) + \mathcal{O}(h^3), x \in T. \tag{2.5}$$

Remark 2.1.1. *Note that the error in formula (2.3) is of first order due to the interpolation of the piecewise constant discontinuous function $m(T), x \in T$, by a continuous function g. This interpolation is mutatis mutandis the interpolation of a continuous function by a piecewise constant function the error of which is known to be of first order. Thus, it is impossible to improve the order of the error term in (2.3).*

In the special case $g \equiv 1$ which Liao investigates, the monitor function f determines *the relative growth or shrinkage of the elements with respect to the previous mesh*, i.e. the mesh on which the deformation takes place. In Liao's methods, the monitor function f does in general *not describe the absolute distribution of the element size in space*. If and only if the starting mesh has equidistributed element sizes, the monitor function does control the absolute element size for Liao's methods.

In the new method, considering g to be the area distribution on the undeformed mesh, f describes in contrast the absolute mesh size distribution of the target grid. Thus this area distribution is *independent of the starting grid*.

Note that condition (2.2) ensures that boundary points only can move along the boundary.

Algorithm 2.1.2. *Based upon Liao's work, [18, 22, 57, 58], the transformation Φ is computed in four steps.*

1. *Scale the monitor function $f > \varepsilon > 0$ or the area function $g > \varepsilon > 0$ such that*

$$\int_\Omega \frac{1}{f(x)} dx = \int_\Omega \frac{1}{g(x)} dx \qquad (2.6)$$

holds. For convenience, it is assumed that (2.6) is fulfilled from now on. Let $\tilde{f}$ and $\tilde{g}$ denote the reciprocals of the scaled functions f and g.

2. *Compute a grid-velocity vector field $v : \Omega \to \mathbb{R}^2$ satisfying*

$$-\mathrm{div}(v(x)) = \tilde{f}(x) - \tilde{g}(x),\ x \in \Omega, \quad and \quad v(x) \cdot \mathfrak{n} = 0,\ x \in \partial\Omega, \qquad (2.7)$$

with $\mathfrak{n}$ being the outer normal vector of the domain boundary $\partial\Omega$, which may consist of several components. This is done by calculating

$$-\Delta w = \tilde{f} - \tilde{g}, \quad \partial_{\mathfrak{n}} w = 0 \ on \ \partial\Omega. \qquad (2.8)$$

and setting $v := \nabla w$.

3. *Solve for each grid point x the initial value problem (IVP)*

$$\frac{\partial \varphi(x,t)}{\partial t} = \eta(\varphi(x,t),t), \quad 0 \le t \le 1, \quad \varphi(x,0) = x \qquad (2.9)$$

with

$$\eta(y,s) := \frac{v(y)}{s\tilde{f}(y) + (1-s)\tilde{g}(y)}, \quad y \in \Omega, s \in [0,1]. \qquad (2.10)$$

4. *Define*

$$\Phi(x) = \varphi(x,1)$$

By this method, it is possible to construct a mapping Φ which satisfies the conditions (2.1) and (2.2).

Theorem 2.1.3. *Let Ω have a $\mathcal{C}^{3,\alpha}$-smooth boundary, let $f, g \in \mathcal{C}^1(\bar{\Omega})$ be defined as above. Then, if the mapping $\Phi : \Omega \to \Omega$ constructed above exists, it fulfils conditions (2.1) and (2.2).*

Proof. Define the auxiliary function

$$H(x,t) := |J\varphi(x,t)| \left[t\tilde{f}(\varphi(x,t)) + (1-t)\tilde{g}(\varphi(x,t)) \right]. \qquad (2.11)$$

A direct, but tedious calculation (see Appendix) shows that

$$\frac{\partial H(x,t)}{\partial t} = 0, \qquad (2.12)$$

and therefore one obtains

$$
\begin{aligned}
\frac{1}{g(x)} = \tilde{g}(x) &= |J\varphi(x,0)|\, \tilde{g}(\varphi(x,0)) \\
&= H(x,0) \\
&\overset{(2.12)}{=} H(x,1) \\
&= |J\varphi(x,1)|\, \tilde{f}(\varphi(x,1)) \\
&= |J\Phi(x)|\, \frac{1}{f(\Phi(x))}
\end{aligned}
$$

and by this the proof is finished. $\qquad\qquad\square$

The existence of such a mapping Φ is guaranteed by the following theorem by Dacorogna and Moser [35]. However, the mapping Φ is *not unique*.

Theorem 2.1.4 (Moser)**.** *Let $0 \geq k \in \mathbb{N}$, $\alpha > 0$. Let $\Omega \subset \mathbb{R}^n$ be a domain with $\mathcal{C}^{3+k,\alpha}$-smooth boundary. Suppose $f, g \in \mathcal{C}^{k,\alpha}(\bar{\Omega})$ with $f, g > 0$ in $\bar{\Omega}$ and $\int_\Omega 1/f = \int_\Omega 1/g$. Then, there exists a $\mathcal{C}^{k+1,\alpha}$-diffeomorphism $\Phi : \bar{\Omega} \to \mathbb{R}^n$, which fulfils*

$$g(x)|J\Phi(x)| = f(\Phi(x)) \quad \forall x \in \Omega$$

and

$$\Phi(x) = x \quad \forall x \in \partial\Omega.$$

2.2 Context of the new grid deformation method

Many of the physical and technical phenomena to be simulated numerically are time-dependent problems. Therefore, it seems desirable to be able to use time-dependent monitor functions, as the regions of the domain to refine and to coarsen the grid may change in time. Liao and his coworkers developed such a method [18, 22, 58], which will be called Liao's dynamic method here. It is now described shortly. For the sake of simplicity, we assume that the domain itself does not change over time.

Given a time-dependent monitor function $f(x, t) > 0$ with

$$\int_{\Omega(t)} \frac{1}{f(x, t)} dx = |\Omega|, \quad t_0 < t < T,$$

then a time-dependent bijective transformation $\Phi : \Omega \to \Omega$ is desired which fulfils

$$|J\Phi(x, t)| = f(\Phi(x, t), t), \quad \Phi(x, t) \in \partial\Omega \quad \forall x \in \partial\Omega. \tag{2.13}$$

Φ is calculated in 2 steps:

1. Compute a vector field $v(x, t)$ with

$$-\text{div}(v(x, t)) = \frac{\partial}{\partial t}\left(\frac{1}{f(x, t)}\right), x \in \Omega \quad \text{and } v(x, t) = g(x, t), x \in \partial\Omega. \tag{2.14}$$

   Here, g denotes a vector field defining the movement of the boundary. As an alternative, one can choose $v(x, t) \cdot \mathfrak{n} = 0$ as boundary condition.

2. Solve the IVP

$$\frac{\partial}{\partial t}\Phi(x, t) = f(\Phi(x, t), t) \cdot v(\Phi(x, t), t), \quad t_0 \le t \le T, \quad \Phi(x, 0) = x. \tag{2.15}$$

The enhanced deformation method described in the previous section is in this context a special case of Liao's dynamic method: Consider a static monitor function $f_s(x)$. It is necessary now to construct a time-dependent monitor function f_d, as by using f_s directly, the right hand side of the divergence equation (2.14) would be zero and therefore no deformation would occur. Using the abbreviations $\tilde{f}_s$ and $\tilde{g}$ from above, f_d is defined by

$$\frac{1}{f_d(x, t)} := t\tilde{f}_s(x) + (1 - t)\tilde{g}(x), \quad 0 \le t \le T := 1.$$

For the right hand side of (2.14), one obtains

$$\frac{\partial}{\partial t}\left(\frac{1}{f_d(x)}\right) = \tilde{f}_s(x) - \tilde{g}(x)$$

and, inserting this, one ends up with the divergence equation (2.7) of the static method. Note that for the special choice of f_d, the vector field v is constant in time. Inserting the special monitor function f_d into the ODE (2.15) of the dynamic method, one achieves

$$\frac{\partial\Phi(x, t)}{\partial t} = f_d(\Phi(x, t), t) \cdot v(\Phi(x, t), t) = \frac{v(\Phi(x, t))}{t\tilde{f}_s(\Phi(x, t)) + (1 - t)\tilde{g}(\Phi(x, t))}, 0 \le t \le 1.$$

This is the ODE of the static method, and due to the same initial condition $\Phi(x,0) = x$, the IVPs of both methods coincide. Therefore, at $t = 1$, the enhanced deformation method and Liao's dynamic method with the monitor function f_d and the boundary condition $v(x,t)\cdot\mathbf{n} = 0$ end up with the very same mesh.

Liao's dynamic method itself is a special case of the GCL method proposed by Huang and Russell [24]. For the sake of simplicity, the deformed coordinates will be denoted by X, $X = \Phi(x,t)$. The starting point for this method is the *geometric conservation law (GCL)*, which states that for an arbitrary subdomain $D \subseteq \Omega$

$$\frac{d}{dt} \int_{\Phi(D,t)} dX = \int_{\partial\Phi(D,t)} \frac{\partial}{\partial t}\Phi(s,t) \cdot \mathbf{n} \, ds$$

holds. Using the divergence theorem and imposing condition (2.13), one obtains

$$\mathrm{div}\,(\tilde{f}(X,t)\, X) = -\frac{\partial}{\partial t}\tilde{f}(X,t), \tag{2.16}$$

where $\tilde{f}$ denotes as above the reciprocal of the monitor function. The monitor function has to fulfil the compatibility condition

$$\frac{d}{dt} \int_\Omega \frac{1}{f(x,t)} \, dx = 0 \quad \forall t. \tag{2.17}$$

Here and in the following, the ∇-Operator as well as the div and curl-operator are with respect to the physical (and unknown) coordinates X. By Helmholtz' theorem, every smooth vector field ϕ can be decomposed into an orthogonal sum of a gradient of a scalar field and the curl of another vector field. As for every smooth ϕ the relation div(curlϕ) $\equiv 0$ holds, it is possible to add the curl of an arbitrary but smooth vector field to a solution of (2.16) without affecting its fulfilment. Therefore, a side condition must be stated to provide the uniqueness of the vector field v. In [24],

$$\mathrm{curl}\,(\mu(X,t) \cdot (v(X,t) - \rho(X,t))) = 0 \tag{2.18}$$

is specified, where μ denotes a (scalar) positive time-dependent weighting function and ρ an underlying time-dependent mesh velocity vector field. If equation (2.18) holds, there is a potential w with

$$v(X,t) = \rho(X,t) + \frac{1}{\mu(X,t)}\nabla w(X,t) \quad \forall X \in \Omega. \tag{2.19}$$

Inserting this equation into (2.16) and into the boundary condition $v \cdot \mathbf{n} = 0$ on $\partial\Omega$, one gets the PDE system

$$\begin{cases} \mathrm{div}\left(\frac{\tilde{f}}{\mu}\nabla w\right) = -\frac{\partial}{\partial t}\tilde{f} - \mathrm{div}(\tilde{f}\,\rho) & \forall X \in \Omega \\ \frac{\partial}{\partial\mathbf{n}}w = -\mu\rho \cdot \mathbf{n} & \forall X \in \partial\Omega \, . \end{cases} \tag{2.20}$$

Note that the vector field is not computed in the computational coordinates x, but in the (unknown) deformed coordinates $\Phi(x,t)$. To obtain the desired transformation $\Phi(x,t)$, note that by definition $v(x,t) = X_t(x,t) = \Phi_t(x,t)$ holds. Using this formula in eq. (2.19) leads to the ODE

$$\frac{\partial}{\partial t}\Phi(x,t) = \rho(\Phi(x,t),t) + \frac{\nabla w(\Phi(x,t),t)}{\mu(\Phi(x,t),t)}.$$

According to [24], if the compatibility condition (2.17) holds, every minimiser of the functional

$$I[v] = \frac{1}{2}\int_\Omega \left[\left| -\text{div}(\tilde{f}v) + \frac{\partial}{\partial t}\tilde{f} \right|^2 + \left(\frac{\tilde{f}}{\mu}\right)^2 |\text{curl}\,(\mu(v-\rho))|^2 \right] dx \qquad (2.21)$$

fulfils (2.16) and (2.18) . As the functional (2.21) is quadratic and bounded from below by 0, the minimiser is unique. On the other hand, obviously a solution of (2.16) and (2.18) minimises (2.21), and therefore the vector field v is determined uniquely by equations (2.16) and (2.18).

Remark 2.2.1. *Setting $\rho = 0$ and $\mu = \tilde{f}$ in formulation (2.20), Liao's dynamic method is recovered. In the case of a monitor function $\tilde{f}_d(x,t) = t\tilde{f}_s + (1-t)\tilde{g}$, the enhanced Liao's method is recovered. Because of the compatibility condition (2.6), eq. (2.17) is fulfilled intrinsically for f_d.*

Remark 2.2.2 (Uniqueness). *Although the defining equation (2.16) is not sufficient to provide a unique determination of v, the assumption of having a potential w provides uniqueness, as the existence of a potential requires that eq. (2.18) holds. Therefore, by choosing formulation (2.20), the curl of the vector field is implicitly determined by*

$$\text{curl}(f_d(X,t)v(X,t)) = 0$$

and therefore no other side condition is necessary.

An easy consequence of the facts shown above is the following corollary.

Corollary 2.2.3. *The enhanced Liao's method provides a unique solution.*

Recognising Liao's dynamic method and therefore the enhanced Liao's method as special cases of the GCL method, following the work by Cao, Huang and Russell [24] several reformulations of the enhanced deformation method are at hand. When it is not necessary, it is not distinguished between x and X any more. The monitor function $\tilde{f}$ and the area function $\tilde{g}$ may fulfil the compatibility condition (2.6).

1. **direct formulation**

$$
\begin{cases}
-f_d(X,t)\nabla\left[\operatorname{div}\left(f_d(X,t)\frac{\partial}{\partial t}X\right)+\tilde{f}(X)-\tilde{g}(X)\right] \\
\quad +f_d(X,t)\operatorname{curl}\left(\operatorname{curl}\left(f_d(X,t)\frac{\partial}{\partial t}X\right)\right)=0 & \forall X\in\Omega \\
\frac{\partial}{\partial t}X\cdot\mathbf{n}=0 & \forall X\in\partial\Omega \\
\left[\operatorname{curl}\left(\tilde{f}(X)\frac{\partial}{\partial t}X\right)\right]\times n=0 & \forall X\in\partial\Omega
\end{cases}
$$

2. **minimisation formulation**

$$
\begin{cases}
v:=\min_\nu \frac{1}{2}\int_\Omega\left[\;\left|\operatorname{div}(f_d\nu)+\tilde{f}-\tilde{g}\right|^2\right. \\
\qquad\qquad\qquad\left. +\left|\operatorname{curl}(f_d\nu)\right|^2\right]dx & \forall\nu:\nu(x)\cdot\mathbf{n}=0\,\forall x\in\partial\Omega \\
v(x,t)\cdot\mathbf{n}=0 & \forall x\in\partial\Omega \\
\frac{\partial}{\partial t}\Phi(x,t)=v(\Phi(x,t),t) & \forall x\in\Omega
\end{cases}
$$

3. **div-curl formulation**

$$
\begin{cases}
-\operatorname{div}(f_d(x,t)v(x,t))=\tilde{f}(x)-\tilde{g}(x) & \forall x\in\Omega \\
\operatorname{curl}(f_d(x,t)v(x,t))=0 & \forall x\in\Omega \\
\frac{\partial}{\partial t}\Phi(x,t)=v(\Phi(x,t),t) & \forall x\in\Omega \\
v(x,t)\cdot\mathbf{n}=0 & \forall x\in\partial\Omega
\end{cases}
$$

In some publications about grid deformation techniques, a mapping Φ is constructed which fulfils the conditions

$$
\Phi:\partial\Omega\to\partial\Omega \tag{2.22}
$$

as well as

$$
g(\Phi(x))|J\Phi(x)|=f(x)\quad\forall x\in\Omega \tag{2.23}
$$

instead of the similar condition (2.23) as in the enhanced Liao's method. In the following, we present an alternative deformation method which constructs a bijective mapping with the properties (2.2) and (2.23). Like in the case of the enhanced Liao's method, the transformation Φ is computed in four steps.

1. Scale the strictly positive monitor function $f\in\mathcal{C}^1(\bar{\Omega})$ and the strictly positive area distribution $g\in\mathcal{C}^1(\bar{\Omega})$ such that

$$
\int_\Omega f(x)dx=\int_\Omega g(x)dx \tag{2.24}
$$

holds. From now on, we will assume that f and g satisfy eq. (2.24).

2. Compute a velocity vector field $v:\Omega\to\mathbb{R}^2$ which fulfils

$$
\operatorname{div}(v(x))=f(x)-g(x),x\in\Omega,\quad\text{and}\quad v(x)\cdot\mathbf{n}=0,x\in\partial\Omega \tag{2.25}
$$

3. Solve for each grid point x the initial value problem

$$\frac{\partial\varphi(x,t)}{\partial t} = \eta(\varphi(x,t),t), \quad 0 \leq t \leq 1, \quad \varphi(x,0) = x \qquad (2.26)$$

with

$$\eta(y,s) = \frac{v(y)}{sg(y) + (1-s)f(y)}, \quad y \in \Omega, s \in [0,1].$$

4. Define $\Phi(x) := \varphi(x,1)$.

Theorem 2.2.4. *Let f,g be defined like in theorem 2.1.3. If the mapping $\Phi : \Omega \rightarrow \Omega$ constructed by our alternative deformation method exists, equations (2.22) and (2.23) are fulfilled.*

Proof. The proof is based upon the proof of theorem 2.1.3. As f is strictly positive on Ω, its reciprocal $1/f$ is bounded and continuous. The same applies to the area distribution. Therefore, these reciprocals meet the requirements for monitor function and area distribution in the enhanced Liao's method. This method applied to the monitor function $1/g$ and the area distribution $1/f$ constructs a transformation Φ which according to theorem 2.1.3 fulfils

$$\frac{1}{f(x)} J\Phi(x) = \frac{1}{g(\Phi(x))}$$

which is equivalent to (2.23). The algorithm proposed above is gained by replacing f and g in the enhanced Liao's method accordingly. $\qquad \square$

Corollary 2.2.5. *If the deformation vector field v in (2.25) is computed by $v = \nabla w$ with*

$$\Delta w = f - g, \quad \partial_{\mathbf{n}} w = 0 \text{ on } \partial\Omega,$$

the solution of the alternative deformation method is unique.

Remark 2.2.6. *Semper and Liao investigate this method for the special case $g \equiv 1$ [58].*

Remark 2.2.7. *A transformation satisfying (2.23) instead of (2.1) faces the major drawback, that it is not possible to control the distribution of the grid size on the resulting grid. This is due to the fact that the coordinates on the deformed grid are defined by the images of the initial coordinates with respect to Φ. Therefore, the area distribution on the deformed mesh can be expressed as $\mathrm{area}(\Phi(x))$ and thus must be controlled by prescribing $f(\Phi(x))$ rather than $f(x)$. For this reason, we will not investigate the latter deformation method more in detail in this thesis.*

2.3 Numerical realisation of the new deformation method

This section is devoted to the numerical implementation of the four steps of the algorithm 2.1.2 described in the preceding chapter. These steps are:

1. Create and scale the monitor function,

2. solve $-\Delta w = \tilde{f} - \tilde{g}, \quad \partial_n w(x) = 0, x \in \partial\Omega$,

3. solve $\forall x \in \mathcal{V}: \quad \varphi_t(x,t) = \eta(\varphi(x,t),t), \quad \varphi(x,0) = x, \quad 0 \leq t \leq 1$,

4. $\Phi(x) := \varphi(x,1)$.

Although the construction of the mapping Φ can be performed in any dimension, we restrict ourselves to the two-dimensional case. For implementations and investigations of our new grid deformation algorithm in three dimensions, we refer to the Panduranga's master thesis [63]. To prove the existence of a smooth diffeomorphism Φ, Moser uses the smoothness conditions of f, g and the domain Ω stated above. In practical computations, we relax these conditions. In the examples presented in this thesis, the domain Ω may have a Lipschitz boundary. The monitor function f and the area distribution function g however are supposed to be at least Lipschitz-continuous, i.e. Hölder-continuous with $\alpha = 1$.

The first step in constructing Φ is to obtain the functions f and g. If an analytical monitor function f is chosen, i.e. a monitor function defined by a closed analytical formula, the monitor function $\hat{f}$ used in practical computations is defined to be the appropriately scaled bilinear interpolant of f on the current grid. This simplifies the implementation and does not have significant impact to the results, as we will demonstrate in the next section. To compute g, we first determine the element size in a grid point, which is set as the arithmetic mean of the area of the elements surrounding it. Then, we define g as the bilinear interpolant of these node values. Note that the area distribution defined this way is Lipschitz-continuous on the current domain.

The vector field v is computed by solving the pure Neumann problem (2.8) and setting $v := \nabla w$. To maintain a high degree of flexibility with respect to the underlying mesh, we compute problem (2.8) using its discrete weak formulation

$$(\nabla w_h, \nabla \varphi_h) = (\tilde{f} - \tilde{g}, \varphi_h) \quad \forall \varphi_h \in \mathcal{Q}_1(\mathcal{T}) \tag{2.27}$$

by the Finite Element Method on the given mesh $\mathcal{T}$. In the following, every FEM calculation is performed with bilinear conforming Finite Elements unless stated differently.

The solution of the corresponding algebraic systems however requires special care, as the solution of the Neumann problem (2.27) is unique up to an additive constant only. Therefore, we use a modified multigrid method in which after every iteration the side condition $\int_\Omega w_h(x)\,dx = 0$ is imposed by adding a suitable constant to the solution.

Afterwards, the vector field v is approximated by the recovered gradient v_h of the Finite Element solution w_h. For the reconstruction of the gradient we employ superconvergent gradient recovery techniques (cf. [83, 87]) as well as standard 1st-order interpolation techniques. The influence of the recovery method on accuracy and robustness of the deformation algorithm will be investigated more in detail in the subsequent chapter. Due to the gradient recovery employed, v_h is a function belonging to $\mathcal{Q}_1(\mathcal{T})$ and is thus continuous.

In the next step, we approximate the initial value problem (2.9) by replacing v by its discrete counterpart v_h. This leads to the initial value problem

$$\frac{\partial \varphi_h(x,t)}{\partial t} = \eta_h(\varphi_h(x,t),t), \quad 0 \le t \le 1, \quad \varphi_h(x,0) = x \qquad (2.28)$$

with

$$\eta_h(y,s) := \frac{v_h(y)}{s\tilde{f}(y) + (1-s)\tilde{g}(y)}, \quad y \in \Omega, s \in [0,1].$$

Note that this ODE system decouples into 1D-ODEs for every coordinate of every grid node, which can be solved by standard ODE methods. All numerical solution methods for ODEs require one or more evaluations of the right hand side per time step. In the proposed grid deformation method, evaluating the vector field as well as $\tilde{f}$ and $\tilde{g}$ is rather expensive, as every evaluation requires searching through the grid: To evaluate a Finite Element function in a given point in real coordinates, the element this point belongs to has to be known. Unfortunately, we have to evaluate (from the 2nd time step on) at points which possibly have been moved to an entirely different region of the grid during the time steps already performed. Hence, the new element the moved point is in has to be found. Note that the evaluations of v_h, $\tilde{f}$ and $\tilde{g}$ have to take place on the original grid, where these functions have been computed. It is important to realise that the grid deformation process described here needs to search the whole grid (at least one time) *per time step* and *per grid point*. This indicates that the efficient solution of the ODEs requires efficient searching strategies, as the computational effort for searching is likely to be the dominating part in terms of overall runtime. Consequently, searching the grid just by "brute force", i.e. without clever search algorithms, leads to unreasonable computational costs (compare example 2.3.2) with quadratic complexity, as we will show.

To perform the "brute force" approach, for each point we loop over all elements in the grid. The search is finished and the loop terminated if the element containing the point is found. To improve this strategy, we introduce the "improved brute force" approach: Here, we store the index of the element the point has been inside of in the previous time step. If the point is not inside T_{old} after the current time step, we perform a "brute force" search. For both search methods, the whole grid has to be searched in the worst case. Therefore, for a grid consisting of N grid points the search time is $\mathcal{O}(N \cdot NEL) = \mathcal{O}(N^2)$. Note

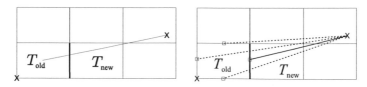

Figure 2.3.1: The principles of raytracing and distance search

that the number of elements *NEL* and the number of vertices N grow with the same order, i.e. $N = \Theta(NEL)$.

Our raytracing search adopted from computer graphics avoids searching the entire grid. At first, it is tested if the grid point is inside the element T_{old} where it was in the previous time step. If this is true, the search is finished. Otherwise, we take the center of T_{old} and connect it with the moved point. Detecting the element edge e which intersects with this ray, we move to the element T_{new} which shares e with T_{old}. Then, setting $T_{old} := T_{new}$, we proceed as before (compare figure 2.3.1, left).

Special care, however, must be taken to avoid that the ray intersects the element in an element vertex as the choice of e is not unique in this case. If this occurs, we simply modify the starting point of the ray heuristically by disturbing the coordinates of the element center.

Another case requiring special treatment may occur when searching in non-convex domains. It may happen that, even though the grid point before and after the current time step is located in the domain, the ray between these points leaves the domain. In this case, we find the second intersection of this ray with the domain boundary by looping over all elements containing at least one boundary edge. In the element containing this second intersection, the search is continued.

As an alternative to the raytracing search, we propose distance search. Again, we first test if the grid point is still inside the element T_{old} where it was in the previous time step. If this is not the case, we compute the squares of the distances of the edge midpoints of T_{old} to the point we search for. As new element T_{new}, we take the one which is adjacent to the edge with the shortest distance (cf. figure 2.3.1, right). However, it may happen that for special configurations the distance information leads to a wrong search direction and thus the search may trap into an infinite loop (compare figure 2.3.2): In the presented example, the comparison of the distances leads to the choice of T_{false} as next candidate. But on this element, the comparison of the distances leads back of T_{old} as successor. Therefore, if during the search T_{new} is chosen as the previous element, we declare the distance search failed and start a raytracing search on T_{old} as fallback. As one search step in the distance search needs less arithmetic operations than one search steps of the raytracing search, it seems to be reasonable that distance search will perform

Figure 2.3.2: In this situation, distance search fails

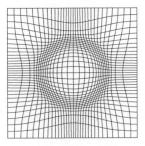

Figure 2.3.3: Resulting grid for test problem 2.3.1, 1,024 elements

faster than its raytracing counterpart.

The maximum length of the search path, i.e. the number of elements touched during the search process, and therefore the maximum amount for a single search process in both raytracing and distance search is $\mathcal{O}(N)$. Thus, the total search time behaves like $\mathcal{O}(N^2)$ in the worst case like for brute force search. But on reasonable grids which have roughly the same resolution in x- and y-direction, the maximum length of the search path is $\mathcal{O}(\sqrt{N})$ and thus the total search time grows like $\mathcal{O}(N^{3/2})$ in contrast to the brute force approaches, where the search time still shows quadratic growth.

Test Problem 2.3.1. *As a first test problem, we consider the unit square $\Omega = [0,1]^2$ triangulated by an equidistant tensor product mesh. We define the monitor function f setting*

$$f(x) = \min\left\{1, \max\left\{\frac{|d - 0.25|}{0.25}, \varepsilon\right\}\right\}, \quad d := \sqrt{(x_1 - 0.5)^2 + (x_2 - 0.5)^2}.$$

(2.29)

The parameter ε is set to 0.1. This choice implies that on the deformed grid the largest cell has 10 times the area of the smallest one. In figure 2.3.3, we show a resulting grid consisting of 1,024 elements.

Example 2.3.2. *To evaluate the time spent with searching the grid during the solution of the ODE (2.28), we consider test problem 2.3.1. The code is compiled using the Intel Fortran Compiler v. 9.1 with full optimisation and runs on a Opteron 250 server equipped with a 64-bit Linux operating system. In this example,*

NEL	distance	β	raytracing	β	imp. br. f.	β	brute force	β
256	$5.15 \cdot 10^{-3}$	-	$5.40 \cdot 10^{-3}$	-	$2.44 \cdot 10^{-2}$	-	$1.34 \cdot 10^{-1}$	-
1,024	$2.27 \cdot 10^{-2}$	4.41	$2.20 \cdot 10^{-2}$	4.07	$6.30 \cdot 10^{-1}$	25.8	$2.13 \cdot 10^{0}$	15.9
4,096	$9.48 \cdot 10^{-2}$	4.17	$9.52 \cdot 10^{-2}$	4.32	$1.87 \cdot 10^{1}$	29.7	$3.53 \cdot 10^{1}$	16.6
16,384	$4.27 \cdot 10^{-1}$	4.50	$4.55 \cdot 10^{-1}$	4.78	$4.06 \cdot 10^{2}$	21.7	$5.59 \cdot 10^{2}$	15.8
65,536	$2.05 \cdot 10^{0}$	4.80	$2.38 \cdot 10^{0}$	5.23	$7.49 \cdot 10^{3}$	18.4	$9.79 \cdot 10^{3}$	17.5
262,144	$1.12 \cdot 10^{1}$	5.46	$1.41 \cdot 10^{1}$	5.92	-	-	-	-
1,048,576	$6.77 \cdot 10^{1}$	6.05	$9.33 \cdot 10^{1}$	6.61	-	-	-	-

Table 2.3.1: Total CPU search time measured in seconds needed from the different search methods and growth factors β in the case of test problem 2.3.1, 10 time steps

NEL	distance	β	raytracing	β	imp. br. f.	β	brute force	β
256	$2.50 \cdot 10^{-2}$	-	$2.49 \cdot 10^{-2}$	-	$4.87 \cdot 10^{-2}$	-	$6.62 \cdot 10^{-1}$	-
1,024	$9.77 \cdot 10^{-2}$	3.91	$1.00 \cdot 10^{-1}$	4.02	$7.97 \cdot 10^{-1}$	16.4	$1.11 \cdot 10^{1}$	16.7
4,096	$3.95 \cdot 10^{-1}$	4.04	$4.06 \cdot 10^{-1}$	4.06	$2.45 \cdot 10^{1}$	30.7	$1.83 \cdot 10^{2}$	16.5
16,384	$1.64 \cdot 10^{0}$	4.15	$1.69 \cdot 10^{0}$	4.16	$7.19 \cdot 10^{2}$	29.3	$2.81 \cdot 10^{3}$	15.4
65,536	$7.00 \cdot 10^{0}$	4.27	$7.35 \cdot 10^{0}$	4.35	$2.05 \cdot 10^{4}$	28.5	$4.50 \cdot 10^{4}$	16.0
262,144	$3.14 \cdot 10^{1}$	4.49	$3.51 \cdot 10^{1}$	4.78	$4.94 \cdot 10^{5}$	24.1	-	-
1,048,576	$1.53 \cdot 10^{2}$	4.87	$1.83 \cdot 10^{2}$	5.21	-	-	-	-

Table 2.3.2: Total CPU search time measured in seconds needed from the different search methods and growth factors β in the case of test problem 2.3.1, 50 time steps

10 time steps with step size Δt fixed to 0.1 are performed using the Runge-Kutta 3 method. Being an three-stage Runge-Kutta method, it requires three evaluations per ODE time step and thus yields 60 evaluations per grid point. For details regarding the ODE solver, we refer to section 3.1. We compare the total search time in seconds, i.e. the search time needed for all evaluations performed in the deformation process, of the four search algorithms described above on various levels of (regular) refinement. Due to the natural inaccuracy in the measurement of time, all measurements have been repeated until the relative error of the mean value of the measured times decreased below 1%.

The search times as well as their growth factors β per regular refinement are collected in table 2.3.1. Tests skipped due to overly long computational times are marked with "-". It is obvious that both distance and raytracing search are by far faster than the brute force approach, which by its quadratic search time (indicated by $\beta \approx 16$) leads to inappropriate search times ($\approx$ 3 h in our example) for still reasonably small grids. The improved brute force search, which is considerably faster than the brute force search on low levels, does not exhibit any significant improvement on fine grids over the brute force search. This indicates that search methods relying on brute force search will lead to unsuitable search times regardless of heuristical improvements. In contrast to this, the search times of distance and raytracing search grow smaller than $\mathcal{O}(N^{3/2})$, which would be indicated by $\beta = 8$. However, the growth factors increase with grid refinement such that it is likely that the search time will grow by a factor of eight when refining a suitable fine grid once more. On very fine grids, where the search paths are longer than on coarse ones, the benefits of the distance search become visible: on the grid consisting of roughly 1 million elements, distance search is roughly 30% faster than the raytracing approach. The same calculations were additionally performed with 50 instead of 10 Runge-Kutta-3 steps and a fixed step size of 0.02. The comparison of the search times in this case (table 2.3.2) with the ones discussed above shows that the total search times for the brute force approaches grows by a factor of 5 corresponding to the number of time steps as expected. Instead of this, the search time in the case of raytracing as well as distance search grows only by a factor of 3 for the calculation with 65,536 elements. This phenomenon stems from the smaller time steps in latter calculations resulting in shorter search paths.

Potentially, hierarchical search methods based on spatial partitions feature even higher speed, as the search time per evaluation needs only $\mathcal{O}(\log N)$ operations yielding a total search time of $\mathcal{O}(N \log N)$. However, these methods require sufficiently long search paths to perform superior because of their inherent overhead of constructing the search hierarchies. Here, the length of the search path is defined by the number of element changes during one search. Consequently, the search path has length zero, if the search ends in the same element where it has started. For test problem 2.3.1, we present in table 2.3.3 the average length of the search paths for searching the grid points during deformation as well as the percentage of searches γ which ended in the same element where they started. Note, that the searches of the intermediate points occurring in the Runge-Kutta method as well as searches of boundary points are not taken into account. As expected, the results indicate that the average search path length doubles in every refinement step. That arises from the fact that a grid point on a fine grid is moved (almost) to the same coordinates than on a rather coarse grid, but the element size is cut into half per refinement, and therefore, the number of elements passed during search doubles. The ratio γ of searches with search path length zero decreases during refinement, which can be explained by the fact that on fine grids small geometric changes result in element changes in contrast to coarse grids, where the elements are large. It is important to note, that even on very fine grids, the search path length is rather small being in the range of 10. This shows that, at least for the example presented here, *the search paths are too small to expect significant speed-ups using hierarchical searching even on very fine grids*. Therefore, we will not investigate hierarchical methods any further in this thesis.

Because of the findings presented, we conduct that both the raytracing and the distance search are well suited searching methods in grid deformation.

Remark 2.3.3. *Table 2.3.3 demonstrates that – as expected – the search path*

	$\Delta t = 0.1$			$\Delta t = 0.02$		
NEL	avg. path length	β	γ in %	avg. path length	β	γ in %
256	$1.87 \cdot 10^{-1}$	-	85.1	$3.61 \cdot 10^{-2}$	-	97.0
1,024	$3.64 \cdot 10^{-1}$	1.95	70.1	$7.13 \cdot 10^{-2}$	1.98	93.7
4,096	$7.09 \cdot 10^{-1}$	1.95	46.2	$1.42 \cdot 10^{-1}$	1.99	87.1
16,384	$1.42 \cdot 10^{0}$	2.00	25.4	$2.83 \cdot 10^{-1}$	1.99	75.1
65,536	$2.82 \cdot 10^{0}$	1.99	12.7	$5.63 \cdot 10^{-1}$	1.99	54.5
262,144	$5.62 \cdot 10^{1}$	1.99	5.64	$1.13 \cdot 10^{0}$	2.01	31.5
1,048,576	$1.12 \cdot 10^{1}$	1.99	2.58	$2.25 \cdot 10^{0}$	1.99	17.1

Table 2.3.3: Average search path length in the case of test problem 2.3.1, 10 and 50 RK3-time steps, respectively

length doubles per refinement step and strongly depends on the number of time steps. Therefore, it seems to be evident to double the number of time steps per grid refinement. Doing so, we can expect the average search path length to remain constant with respect to the grid refinement level. However, in this case the total number of searches during deformation grows like $\mathcal{O}(N^{3/2})$ such that the total searchtime again behaves like $\mathcal{O}(N^{3/2})$. We will come back to these considerations in much more detail in the subsequent chapter 3 in the context of convergence investigations for our grid deformation method.

Remark 2.3.4. *In three dimensions, the advantages of hierarchical searching methods will be even less pronounced than in two dimensions, as on grids emerging from regular refinement, the search path length grows like $\mathcal{O}(N^{4/3})$ instead of $\mathcal{O}(N^{3/2})$.*

In this section, we developed and tested the numerical components we need to perform grid deformation by computing algorithm 2.1.2 numerically. To summarise this section, we give a short description of our basic grid deformation method in algorithmic form.

Algorithm 2.3.5 (Basic grid deformation).

input: • f: *monitor function*
           • *GRID: computational grid*

output: • *GRID: deformed grid*

function Deformation(f, $GRID$) : $GRID$

   compute $\tilde{f} - \tilde{g}$, $\tilde{g} = \tilde{g}(GRID)$

   solve $(\nabla w_h, \nabla \varphi_h) = (\tilde{f} - \tilde{g}, \varphi_h) \quad \forall \varphi_h \in \mathcal{Q}_1(\mathcal{T})$

   $v_h :=$ **recovered_gradient** (w_h)

DO FORALL $x \in GRID$

    solve $\frac{\partial \varphi(x,t)}{\partial t} = \eta_h(\varphi(x,t),t), \quad 0 \leq t \leq 1, \quad \varphi(x,0) = x$

    $\Phi(x) := \varphi(x,1)$

ENDDO

RETURN $GRID$

END Deformation

Chapter 3

Theoretical and numerical analysis of the grid deformation algorithm

In this section, we show that the numerical realisation of the new grid deformation algorithm 2.1.2 converges in a certain sense under rather mild conditions. The method is analysed with respect to computational effort, the quality of the resulting grid and the numerical difficulties connected to the discretisation of the deformation method. Furthermore, based upon the experiences gained in these numerical experiments we define and test improved deformation algorithms.

3.1 Accuracy aspects and convergence

To assess the accuracy aspects related to grid deformation, we need a suitable way to measure "accuracy". As our grid deformation algorithm described in the preceding chapter aims to create a grid with a prescribed area distribution, we consider the fraction

$$\frac{f(x)}{\text{area}(x)} \tag{3.1}$$

where area(x) stands for the interpolated element area distribution of the deformed grid. If X is a vertex of the deformed grid, area(X) is defined to be the arithmetic mean value of the areas of the elements X belongs to and by the bilinear interpolation of the corresponding node values otherwise. If the desired area distribution is achieved, then the fraction (3.1) is spatially constant (note that f describes the desired spatial distribution of the cell sizes up to a spatially fixed constant). To force this constant to be 1, we scale the function area(x) with a multiplicative constant c to achieve

$$\int_\Omega a(x)\, dx = \int_\Omega f(x)\, dx, \tag{3.2}$$

where $a(x) := c \cdot \text{area}(x)$. Now, we define the quality function

$$q(x) := \frac{f(x)}{a(x)}.$$

The overall quality of the grid adaptation according to the desired cell size can be measured by the deviation of $q(x)$ from the constant function 1 leading to the quality measures Q_0 and Q_∞ defined by

$$Q_0 := ||q - 1||_0 \quad \text{and} \quad Q_\infty := \max_{x \in \mathcal{V}} |q(x) - 1|. \tag{3.3}$$

Remark 3.1.1. *Note that the quality measures defined in equation (3.3) are consistent with respect to regular refinement, i.e. regular refinement of a deformed grid does not change the quality measures systematically. Let k denote the number of regular refinement steps after the deformation process, let $\text{area}_k(x)$ be the area distribution of the refined grid after the regular refinement and $a_k(x)$ its scaled counterpart. Then, $\text{area}(x) = \mathcal{O}(4^k) \cdot \text{area}_k(x)$ holds, but by virtue of equation (3.2), we end up with $a(x) = \mathcal{O}(1) \cdot a_k(x)$. Therefore, we have $q_k(x) = \mathcal{O}(1) \cdot q(x)$, where q_k stands for the ratio of monitor function f and the scaled area distribution a_k on the refined grid.*

As alternative to our quality measure, one could e.g. consider for a certain grid point x_i the euclidian length of the difference of the coordinates X_i computed with the exact solution of the PDE (2.8) and the IVP (2.9) and the actually computed coordinates $\tilde{X}_i$. The corresponding natural global measure then is

$$(\tilde{Q}_0)^2 := \frac{1}{N} \sum_{i=1}^{N} ||X_i - \tilde{X}_i||^2 = \frac{1}{N} \sum_{i=1}^{N} ||\Phi(x_i) - \tilde{\Phi}(x_i)||^2,$$

where $\tilde{\Phi}$ stands for the approximate transformation computed with numerical methods. In the same way, we define $\tilde{Q}_\infty$ applying the maximum norm. However, there are two drawbacks: In contrast to Q, $\tilde{Q}$ is hard to compute, as the exact transformation Φ is of course unknown. Moreover, the uniqueness of Φ in our deformation method is given by the implicit condition $curl\, v = 0$ to the deformation vector field, which is somewhat artificial. Thus, $\tilde{Q}$ does not take into account that there are in fact infinitely many ways to construct a grid with the desired area distribution and thus there is no unique "reference transformation" Φ we could compare with. Therefore, we only employ Q_0 and Q_∞ in what follows.

In practical computations, one can not expect that the quality measures become arbitrarily small, as the numerical realisation of the deformation algorithm induces numerical errors. Moreover, a consistency error occurs which stems from the fact that the actual cell size distribution is discontinuous and has to be interpolated in order to gain $\text{area}(x)$. Therefore, even when solving the deformation PDE (2.8) and all IVPs (2.9) *exactly*, we cannot expect one of the quality measures to be zero. Overall, there are three main error sources which occur during in the deformation process:

1. The deformation PDE (2.8) is solved approximately.

2. The IVPs (2.9) are solved approximately.

3. The interpolation of the discontinuous cell size distribution induces a *consistency error*.

One key aspect in numerical mathematics is the convergence of the approximate solution gained by a numerical algorithm to an exact one. In the following analysis we consider for a given bounded domain Ω a sequence of triangulations $(\mathcal{T}_i)_{i \in I}$ where I denotes an arbitrary index set. Let us denote the number of vertices in $\mathcal{T}_i$ by N_i. In what follows, we always assume that $\bigcup_{T \in \mathcal{T}_i} T = \Omega \ \forall i \in I$ and $N_{i+1} > N_i \ \forall i \in I$. The following definitions prepare the analysis of the convergence behaviour of our grid deformation to follow in this chapter.

Definition 3.1.2. *a) The sequence of triangulations $(\mathcal{T}_i)_{i \in I}$ is said to be* edge-length regular, *iff*

$$h_i := \max_{e \in \mathcal{E}_i} |e| = \mathcal{O}(N_i^{-0.5}) \quad \forall i \in I.$$

b) The sequence of triangulations $(\mathcal{T}_i)_{i \in I}$ is said to be size regular, *iff*

$$\exists 0 < c, C : ch_i^2 \leq m(T) \leq Ch_i^2 \quad \forall T \in \mathcal{T}_i \ \forall i \in I. \tag{3.4}$$

For the initial grids, we postulate edge-length regularity from now on without explicitly stating this. Note that for an edge-length regular sequence of grids, a convergence analysis in powers of h is justified. This enables the definition of a *similarity condition* for our sequence.

Definition 3.1.3. *An edge-length regular sequence of triangulations $(\mathcal{T}_i)_{i \in I}$ fulfils the* similarity condition, *iff there is a function g with $0 < g_{\min} \leq g \leq g_{\max} < \infty$ and there are positive constants c_s, C_s with*

$$\frac{1}{h_i^2} c_i m(T) = g(x) + \mathcal{O}(h_i) \quad \forall x \in T \quad \forall T \in \mathcal{T}_i \ \forall i \in I, \quad c_s \leq c_i \leq C_s. \tag{3.5}$$

Here, c_i is a spatially fixed constant.

If a sequence of grids fulfils the similarity conditions, all grids feature a similar distribution of element sizes up to a spatially fixed constant. This condition is naturally satisfied for a sequence of grids created by successive regular refinement of an arbitrary coarse grid. Notice that these requirements affect the asymptotic behaviour only and permit large spatial variations of both element size and shape.

For edge-length regular sequences of triangulations, the similarity condition implies size regularity.

Lemma 3.1.4. *Let the sequence of triangulations $(\mathcal{T}_i)_{i \in I}$ fulfil the similarity condition. Then, it is size regular.*

Proof. Consider an arbitrary element $T \in \mathcal{T}_i$. Then,

$$m(T) = \frac{h_i^2}{c_i} g(x) + \underbrace{\mathcal{O}(h_i^3)}_{\leq ch_i^3} \leq \left(\frac{g_{\max}}{c_s} + c \right) h_i^2$$

for $h_i < 1$. On the other hand,

$$m(T) = \underbrace{\frac{h_i^2}{c_i} g(x)}_{>0} + \underbrace{\mathcal{O}(h_i^3)}_{\geq -ch_i^3} \geq \left(\frac{g_{\min}}{C_s} - ch_i \right) h_i^2 \geq \frac{g_{\min}}{2C_s} h_i^2$$

for $h_i < g_{\min}/(cC_s)$. $\qquad\square$

In what follows, the monitor function f is supposed to be fixed. Let us denote the approximate transformation computed on $\mathcal{T}_i$ by $\tilde{\Phi}_i$. With the influence of numerical errors in mind, it seems to be natural to define convergence by $\tilde{\Phi}_i \to \Phi$ in a certain norm. Unfortunately, Φ is unique by the rather artificial condition $curl\,v = 0$ only, which makes the proposed definition of convergence questionable.

Therefore, we define convergence by the decay of our quality measures.

Definition 3.1.5. *For an edge-length regular sequence $(\mathcal{T}_i)_{i \in I}$, a sequence of grid deformations is said to converge, iff $Q \to 0$ for $h \to 0$. Here, Q stands for either Q_0 or Q_∞.*

To formulate the following results, we need some notations. For an arbitrary grid point x, we denote by X its image in the deformed grid obtained with both PDE (2.8) and the initial value problem $\partial_t \varphi(x,t) = \eta(\varphi(x,t),t), \varphi(x,0) = x$ solved exactly. Let us denote by X_h the image of the same vertex x, but computed by solving exactly the disturbed initial value problem $\partial_t \varphi(x,t) = \eta_h(\varphi(x,t),t)$ according to formula (2.28), where the exact vector field v is replaced by v_h. Furthermore, $\tilde{X}$ is the corresponding point obtained by computing both the deformation vector field and the initial value problem numerically. The following two lemmas demonstrate that size and shape regularity are preserved under grid deformation, if our deformation algorithm is carried out solving all differential equations exactly. As an abbreviation we call these grids "exactly deformed", and "numerically deformed" if the differential equations have been solved by numerical methods.

Lemma 3.1.6. *Let $(\mathcal{T}_i)_{i \in I}$ be a sequence of grids with grid size $h_i \to 0$ which fulfils the similarity condition (3.5). Then the sequence $(\mathcal{T}_i^d)_{i \in I}$ of exactly deformed meshes is edge-length regular.*

Proof. Let us denote by X and Y the images of the vertices x and y obtained by exact computation. These vertices are computed by solving the initial value problems

$$\begin{aligned} \partial_t \varphi^x(t) &= \eta(\varphi^x(t),t), \quad \varphi^x(t=0) = x \\ \partial_t \varphi^y(t) &= \eta(\varphi^y(t),t), \quad \varphi^y(t=0) = y \end{aligned}$$

with the right hand side η defined in formula (2.10). According to Gronwall's lemma (compare [75, p. 16]), it holds

$$||\varphi^x(t) - \varphi^y(t)|| \le e^{Lt}(\underbrace{||\varphi^x(0) - \varphi^y(0)||}_{=||x-y||}).$$

Here, the Lipschitz constant L of the ODE right hand side is defined by

$$||\eta(t,x) - \eta(t,x')|| \le L||x - x'|| \, \forall x, x' \in \Omega \, \forall t \in [0,1]. \tag{3.6}$$

Thus, we have

$$
\begin{aligned}
L &\le \sup_{t \in [0,1]} \sup_{x,x' \in \Omega, x \ne x'} \left|\left| \frac{v(x)}{sf(x) + (1-s)g(x)} - \frac{v(x')}{sf(x') + (1-s)g(x')} \right|\right| \, ||x - x'||^{-1} \\
&= \mathcal{O}(h^0). \tag{3.7}
\end{aligned}
$$

From this, we deduce for $t = 1$

$$||\varphi^x(1) - \varphi^y(1)|| = ||X - Y|| \le e^L||x - y|| \le e^L ch.$$

$\square$

Lemma 3.1.7. *Let $f > \varepsilon > 0$ be a strictly positive monitor function, $f \in \mathcal{C}^1(\bar{\Omega})$ and $(\mathcal{T}_i)_{i \in I}$ be a sequence of grids which fulfils the similarity condition (3.5). Then, the sequence $(\mathcal{T}_i^d)_{i \in I}$ of exactly deformed grids is size regular.*

Proof. Because of $f \in \mathcal{C}^1(\bar{\Omega})$, $\exists f_{\max} := \max_{x \in \Omega} |f(x)| < \infty$. By definition 3.1.3, it holds $g_{\min} \le g \le g_{\max}$. Thus,

$$\frac{m(\Phi(T))}{m(T)} \le 2\frac{f_{\max}}{g_{\min}} = \mathcal{O}(h^0)$$

and, vice versa,

$$\frac{m(T)}{m(\Phi(T))} \le 2\frac{g_{\max}}{\varepsilon} = \mathcal{O}(h^0)$$

for $h < h_0$. Therefore, $m(T)$ and $m(\Phi(T))$ have the same convergence order. As the sequence of initial grids is size regular due to lemma 3.1.4, the assertion follows immediately. $\square$

Many sequences of shape- and size regular grids fulfil the condition $|e| \ge ch$ for the length of any edge e. This naturally holds for all grids created by successive regular refinement of an arbitrary coarse grid. Under stricter assumptions than before, this property is preserved by grid deformation as well.

Lemma 3.1.8. *Let $f > \varepsilon > 0$ be a strictly positive monitor function, $f \in \mathcal{C}^1(\bar{\Omega})$ and $(\mathcal{T}_i)_{i \in I}$ be a sequence of grids which fulfils the similarity condition (3.5) and it may hold*

$$\exists c > 0 : |e| \ge ch \quad \forall e \in \mathcal{E}_i \, \forall i \in I. \tag{3.8}$$

Moreover, let the area function in formula (3.5) be smooth, $g \in \mathcal{C}^1(\bar{\Omega})$. If the mapping Φ fulfils the symmetry condition

$$\partial_x \Phi_1(x,y) = \partial_y \Phi_2(x,y) \quad \forall\, (x,y) \in \Omega, \tag{3.9}$$

Φ_j *denoting the j-th component of the vector-valued mapping Φ, there exists $h_0 > 0$ such that*

$$||\Phi(x) - \Phi(x')|| \geq ch \quad \forall e = \overline{xx'} \in \mathcal{E}_i \quad \forall i \in I \text{ with } h_i < h_0.$$

Proof. Under the assumptions made, it holds $\Phi \in \mathcal{C}^2(\bar{\Omega})$ according to theorem 2.1.4. Due to the symmetry condition, the Jacobian $J\Phi$ is symmetric and because of $|J\Phi(x)| = f(\Phi(x))/g(x) > 0$ invertible. Thus, $J\Phi(x)$ has two real eigenvalues $\lambda_{\max}(x) \neq 0$ and $\lambda_{\min}(x) \neq 0$ with $\lambda_{\min}(x)\lambda_{\max}(x) = f(\Phi(x))/g(x)\,\forall\, x \in \Omega$. As $J\Phi \in \mathcal{C}^1(\bar{\Omega})$, the functions $\lambda_{\max}(x)$ and $\lambda_{\min}(x)$ are continuous and thus bounded on the compact set $\bar{\Omega}$. Moreover, $\lambda_{\min}(x)$ and $\lambda_{\max}(x)$ are either both positive or both negative $\forall x \in \Omega$. Otherwise, at least one of these two functions would have at least one zero. Let us assume at first that all eigenvalues are positive. With $\lambda_{\max} := \sup_{x \in \Omega} \lambda_{\max}(x)$ and $g_{\max} := \sup_{x \in \Omega} g(x)$, we have

$$\lambda_{\min}(x) = \frac{f(\Phi(x))}{g(x)\lambda_{\max}(x)} \geq \frac{\varepsilon}{g_{\max}\lambda_{\max}} \qquad \forall\, x \in \Omega.$$

Thus,

$$\lambda_{\min} := \inf_{x \in \Omega} \lambda_{\min}(x) \geq \frac{\varepsilon}{g_{\max}\lambda_{\max}} > 0.$$

Let now x and x' be the start and end point of an arbitrary edge $e \in \mathcal{E}_i$. Then, a Taylor expansion yields

$$||\Phi(x) - \Phi(x')||_2 = ||J\Phi(x')(x - x') + \mathcal{O}(||x - x'||_2^2)||_2$$

with $|| \cdot ||_2$ denoting the standard euclidian norm on $\mathbb{R}^2$. As the last term is of higher order, we obtain

$$2||\Phi(x) - \Phi(x')||_2 \geq ||J\Phi(x')(x - x')||_2 \quad \forall\, h < h_0. \tag{3.10}$$

In the euclidian vector norm, the Rayleigh-quotient is an upper bound for the minimal eigenvalue. Thus, we conclude with equation (3.8)

$$|(J\Phi(x')(x - x'), x - x')| \geq \lambda_{\min}||x - x'||_2^2 > ch^2. \tag{3.11}$$

Let us assume $\exists \alpha > 0 : ||J\Phi(x')(x - x'))|| \leq ch^{1+\alpha}$ for $h < h_0$. Then, according the assumed size and shape regularity,

$$|(J\Phi(x')(x - x'), x - x')| \leq ||J\Phi(x')(x - x')||_2||x - x'||_2 \leq ch^{2+\alpha} \quad \forall h < h_0$$

which contradicts (3.11). This leads to

$$||J\Phi(x')(x - x')||_2 \geq ch, \quad h < h_0,$$

which yields in conjunction with formula (3.10) the assertion.
With the same settings as above, we obtain in the case of negative eigenvalues the equation

$$|(J\Phi(x')(x - x'), x - x')| \geq |\lambda_{\max}| \, ||x - x'||_2^2 > ch^2.$$

which is analogous to formula (3.11). With similar arguments as before, we show

$$|\lambda_{\max}| \geq \frac{\varepsilon}{g_{\max}|\lambda_{\min}|},$$

where the existence of $\lambda_{\min} := \inf_{x \in \Omega} \lambda_{\min}(x)$ is granted by the continuity of $\lambda_{\min}(x)$. Then, we proceed as in the case of positive eigenvalues. $\square$

Combining the lemmata from above, we end up with the following corollary.

Corollary 3.1.9. *Let $f > \varepsilon > 0$ be a strictly positive monitor function, $f \in \mathcal{C}^1(\bar{\Omega})$ and $(\mathcal{T}_i)_{i \in I}$ be a sequence of grids which fulfils the similarity condition (3.5) and condition (3.8). Moreover, for the area function g emerging in the similarity condition (3.5), we require $g \in \mathcal{C}^1(\bar{\Omega})$. If Φ fulfils the symmetry condition (3.9), then there exists $h_0 > 0$ and constants c_1, C_1, c_2 and C_2 such that*

$$c_1 h \leq ||\Phi(x) - \Phi(x')|| \leq C_1 h \quad \forall e = \overline{xx'} \in \mathcal{E}_i \text{ with } h_i < h_0 \forall i \in I.$$

and

$$c_2 h^2 \leq m(\Phi(T)) \leq C_2 h^2 \quad \forall T \in \mathcal{T}_i, i \in I.$$

Now we are able to formulate and prove our central convergence theorem for grid deformation.

Theorem 3.1.10. *Let $(\mathcal{T}_i)_{i \in I}$ be a sequence of grids with grid size $h_i \to 0$ which fulfils condition (3.5) and let us denote the sequence of numerically deformed grids by $(\tilde{\mathcal{T}}_i)_{i \in I}$. For the monitor function f, it may hold $0 < \varepsilon < f \in \mathcal{C}^1(\bar{\Omega})$. Let us assume that for the approximate solution of the deformation vector field v_h, the equation $||v - v_h||_\infty = \mathcal{O}(h^{1+\delta})$, $\delta > 0$ is valid. Let $||X_h - \tilde{X}|| = \mathcal{O}(h^{1+\delta})$ be true for any vertex. Then,*

a) *the sequence of numerically deformed meshes $(\tilde{\mathcal{T}}_i)_{i \in I}$ is edge-length regular,*

b) *$(\tilde{\mathcal{T}}_i)_{i \in I}$ is size regular according to condition (3.4),*

c) *the sequence of deformations converges, moreover, $\exists c > 0$ independent of h, such that*

$$Q_0 \leq ch^{\min\{1,\delta\}}, \quad Q_\infty \leq ch^{\min\{1,\delta\}}. \tag{3.12}$$

42

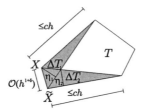

Figure 3.1.1: Influence of an disturbed vertex on the element area

Proof. Let us denote the area distribution on the numerically deformed grid T by $\tilde{a}(x)$. The quantities X_h and X are computed as the solution of the initial value problems

$$
\begin{aligned}
\partial_t \varphi(t) &= \eta(\varphi(t), t), & \varphi(0) &= x, \\
\partial_t \varphi_h(t) &= \eta_h(\varphi_h(t), t), & \varphi_h(0) &= x
\end{aligned}
$$

in $t = 1$. As a consequence of Gronwall's lemma, it holds (compare [75, p. 16])

$$
\begin{aligned}
||X - X_h|| &\leq e^L \int_0^1 \sup_{x \in \Omega} ||\eta(x, s) - \eta_h(x, s)|| \, ds \\
&\leq e^L \int_0^1 \sup_{x \in \Omega} \left|\left| \frac{v(x) - v_h(x)}{sf(x) + (1-s)g(x)} \right|\right| \, ds \\
&\leq e^L \underbrace{\sup_{x \in \Omega} ||v(x) - v_h(x)||}_{=\mathcal{O}(h^{1+\delta})} \underbrace{\sup_{t \in [0,1]} \sup_{x \in \Omega} ||(tf(x) + (1-t)g(x))^{-1}||}_{=\mathcal{O}(h^0)} \\
&= e^L \mathcal{O}(h^{1+\delta}).
\end{aligned}
\tag{3.13}
$$

Here, the Lipschitz constant L is given by the condition (3.6) and fulfils equation (3.7) like in the proof of lemma 3.1.6. The combination of this upper bound with formula (3.13) yields

$$
||X - X_h|| = \mathcal{O}(h^{1+\delta}).
\tag{3.14}
$$

for all grid vertices X. Using the triangle inequality, it follows

$$
||X - \tilde{X}|| \leq \underbrace{||X - X_h||}_{(3.14):\,\mathcal{O}(h^{1+\delta})} + \underbrace{||X_h - \tilde{X}||}_{=\mathcal{O}(h^{1+\delta})} = \mathcal{O}(h^{1+\delta}).
\tag{3.15}
$$

Note that equation (3.15) holds *without any assumptions on initial and deformed mesh*. Let $\tilde{e}$ be an arbitrary edge on the numerically deformed grid with endpoints $\tilde{X}$ and $\tilde{Y}$. Then, due to the triangle inequality and lemma 3.1.6, it follows

$$
||\tilde{X} - \tilde{Y}|| \leq \underbrace{||X - \tilde{X}||}_{(3.15):\,\leq ch^{1+\delta}} + \underbrace{||X - Y||}_{\leq ch} + \underbrace{||Y - \tilde{Y}||}_{(3.15):\,\leq ch^{1+\delta}} \leq ch,
$$

and thus the edge-length regularity of the deformed grid. Let us consider a single element T on the numerically computed deformed grid $\tilde{X}$ is a vertex of. Let us denote by X the exactly computed counterpart of $\tilde{X}$. Then, the area difference ΔT due to the disturbed vertex $\tilde{X}$ can be bounded by $\Delta T \leq \Delta T_1 + \Delta T_2$ (see figure 3.1.1). These quantities are the areas of triangles which basis length is bounded by ch due to the shape regularity. Their heights η_1 and η_2 are not larger than $||X - \tilde{X}||$ and therefore (at most) $\mathcal{O}(h^{1+\delta})$. Because of this, ΔT due to the disturbed vertex $\tilde{X}$ can be bounded by $c_d h^{2+\delta}$. This holds for the other vertices of T as well, so that we can conclude

$$m(\Phi(T)) = m(\tilde{\Phi}(T)) + \mathcal{O}(h^{2+\delta}). \tag{3.16}$$

From this, we get using lemma 3.1.7

$$m(\tilde{\Phi}(T)) \leq \underbrace{m(\Phi(T))}_{\leq Ch^2} + c_d h^{2+\delta} \leq 2Ch^2$$

and vice versa

$$m(\tilde{\Phi}(T)) \geq \underbrace{m(\Phi(T))}_{\geq ch^2} - c_d h^{2+\delta} \geq \frac{1}{2}ch^2$$

for $h < h_0$. This proves assertion b).

By construction, $\tilde{a}(x) = \frac{c}{h^2} m(\tilde{\Phi}(T)) + \mathcal{O}(h)$ holds. By a Taylor expansion of the monitor function f, we obtain

$$\frac{f(\tilde{X})}{\tilde{a}(\tilde{X})} = \frac{f(X) + \nabla f(\chi) \cdot (X - \tilde{X})}{\frac{c}{h^2} m(\tilde{\Phi}(T)) + \mathcal{O}(h)}$$

$$= \frac{f(X)}{\frac{c}{h^2} m(\tilde{\Phi}(T)) + \mathcal{O}(h)} + \underbrace{\frac{\overbrace{\nabla f(\chi) \cdot (X - \tilde{X})}^{=\mathcal{O}(h^{1+\delta})}}{\frac{c}{h^2} m(\tilde{\Phi}(T)) + \mathcal{O}(h)}}_{=\mathcal{O}(h^0)}.$$

Inserting equation (3.16), we obtain

$$\frac{f(\tilde{X})}{\tilde{a}(\tilde{X})} = \frac{f(X)}{\frac{c}{h^2} m(\Phi(T)) + \mathcal{O}(h^{\min\{1,\delta\}})} + \mathcal{O}(h^{1+\delta})$$

Now, we exploit that due to the edge-length regularity of the deformed mesh, the length of any element edge is bounded from above by ch. Let us denote the center

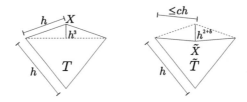

Figure 3.1.2: Sequence of elements which prevents valid grids even if the deformation converges

of the undeformed element by x_c. Then we get by another Taylor expansion

$$
\frac{f(\tilde{X})}{\tilde{a}(\tilde{X})} = \frac{f(\Phi(x_c))}{\frac{c}{h^2}m(\Phi(T)) + \mathcal{O}(h^{\min\{1,\delta\}})} + \overbrace{\underbrace{\frac{\nabla f(\nu) \cdot (\Phi(x_c) - X)}{\frac{c}{h^2}m(\Phi(T)) + \mathcal{O}(h^{\delta})}}_{=\mathcal{O}(h^0)}}^{=\mathcal{O}(h)} + \mathcal{O}(h^{1+\delta})
$$

$$
= \frac{f(\Phi(x_c))}{f(\Phi(x_c))} + \mathcal{O}(h^{\min\{1,\delta\}}) + \mathcal{O}(h)
$$

$$
= 1 + \mathcal{O}(h^{\min\{1,\delta\}}).
$$

As this relation holds for all grid vertices, it follows immediately the assertion $Q_\infty = \mathcal{O}(h^{\min\{1,\delta\}})$. Let now Y be an arbitrary grid point in the interior of the element T with vertices $V_1, \ldots V_4$ on the numerically deformed grid. By $\tilde{X}$, we denote one of the four vertices of T, where the area function $\tilde{a}$ is minimal: $\tilde{a}(\tilde{X}) = \min_{i=1}^{4}\{\tilde{a}(V_i)\}$. As the computed grid is edge-length regular and size regular according to condition (3.4) and due to the normalisation condition (3.2)), we know that $\tilde{a}(\tilde{X}) = \mathcal{O}(h^0)$. Moreover, as $\tilde{a}$ is a bilinear function on T, we have $\tilde{a}(Y) \geq \tilde{a}(\tilde{X}) \, \forall \, Y \in T$. Thus, we gain by a Taylor expansion

$$
\frac{f(Y)}{\tilde{a}(Y)} \leq \frac{f(Y)}{\tilde{a}(\tilde{X})}
$$

$$
= \frac{f(\tilde{X})}{\tilde{a}(\tilde{X})} + \overbrace{\frac{\nabla f(\mu) \cdot (\tilde{X} - Y)}{\tilde{a}(\tilde{X})}}^{\leq ch}
$$

$$
\leq \frac{f(\tilde{X})}{\tilde{a}(\tilde{X})} + ch,
$$

h sufficiently small. This leads to the assertion $Q_0 \leq ch^{\min\{1,\delta\}}$. $\qquad\square$

However, even if we have proven the convergence of our grid deformation, there is no warranty even under the rather strict conditions on the initial grids from theorem 3.1.10 that the deformed grid contains convex elements only, i.e. that

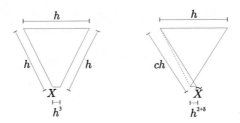

Figure 3.1.3: Another sequence of elements which prevents valid grids even if the deformation converges

0	
1	1
	1/2 1/2

0		
1/2	1/2	
1	−1	2
	1/6	2/3	1/6

0			
1/2	1/2		
1/2	0	1/2	
1	0	0	1
	1/6	1/3	1/3	1/6

Figure 3.1.4: Butcher schemes for the Runge-Kutta methods HEUN (left), RK3 (middle) and RK4 (right)

the deformed grid is admissible. In general, this is not necessarily the case: Let the sequence $(\mathcal{T}_i^d)_{i \in I}$ of deformed meshes contain a sequence of elements T like in figure 3.1.2. It is evident that for the special element shown, computing the vertex X with order $2 + \delta$ is insufficient to guarantee valid meshes but sufficient to obtain convergence. The same holds for the type of element shown in figure 3.1.3. Note that for both elements, the edges are bounded by h and that it holds $m(T) = \mathcal{O}(h^2)$ as required by size- and shape regularity. However, under the assumptions of corollary 3.1.9, this kind of element will not emerge in the deformation because of the property $|e| \geq ch$ proven in the aforementioned corollary.

In the following, we investigate how the error induced by the approximate solution of the IVPs influences the quality measures Q_0 and Q_∞. To do so, we consider test problem 2.3.1 and compare Q_0 and Q_∞ after applying deformation algorithm 2.3.5 on an equidistant tensor product grid. As IVP solver, we apply the explicit Euler's method (EE) and several Runge-Kutta type methods: Heun's

0			
1/4	1/4		
27/40	−189/800	729/800	
	214/891	1/33	650/891

Figure 3.1.5: Butcher scheme for the Runge-Kutta-Fehlberg method RKF2B

method (HEUN), the classical Runge method of third order (RK3) and the standard Runge-Kutta method of fourth order (RK4) (see figure 3.1.4). These convergence orders however, can be experienced only for sufficiently smooth solutions and thus sufficiently smooth right hand sides. The right hand side we consider is continuous only and therefore lacks the required smoothness. Therefore, we can not expect to experience full convergence of the high order methods. Because of this, we additionally investigate the Runge-Kutta-Fehlberg method RKF2B (see [49, p. 167]). For the corresponding Butcher scheme, we refer to figure 3.1.5. This scheme is a three-stage method, but of second order only. Instead of aiming at the highest convergence order possible (three), this method is designed to achieve a particularly small error constant. Furthermore, we employ Adams-Bashforth methods of order two and three (AB2, AB3) as representatives of linear multi-step methods. The starting values are obtained by Heun's method. For all IVP solvers mentioned except the RKF2B method, we refer to lecture notes by Turek and Rannacher [75].

All computations are performed with fixed step size, the deformation vector field is constructed employing the SPR [85, 86, 87, 88] gradient recovery technique. The results are shown in table 3.1.1 as well as in table 3.1.2. Note that one Runge-Kutta step requires several costly evaluations of the right hand side of the ODE per step in contrast to the linear multi-step methods. Therefore, in the corresponding figures 3.1.6 and 3.1.7 we compare Q_0 and Q_∞ in relation to the number of evaluations of the right hand side per grid point instead of the number of time steps. It turns out that the high order Runge-Kutta methods perform best and lead to very precise solutions of the IVPs indicated by low quality measures even when only few large time steps are performed. The numerical results show that for the example presented the error induced by the IVP solve does not have any significant influence if five or more high order Runge-Kutta steps are employed and even using three time-steps only leads to low quality measures in contrast to the linear multi-step methods, where the choice of large time steps causes severe distortions of the grid indicated by large quality measures or even non-convex elements. The RKF2B method can not compete with the high order Runge-Kutta methods RK3 and RK4 in this example. figures 3.1.6 and 3.1.7 reveal that the additional amount of the intermediate evaluations inherent to the Runge-Kutta methods is by far outweighed by the additional robustness and accuracy of these algorithms in comparison to the linear multi-step methods such that given the same number of evaluations of the right hand side, the Runge-Kutta methods provide more accurate results. As the numerical amount in solving the IVPs in this particular situation is dominated by the numerical effort to search the grid and thus by the number of evaluations, we consider the high-order Runge-Kutta methods being best suited for grid deformation purposes. Note that the application of the Runge-Kutta method of fourth order does not lead to significantly reduced quality measures in comparison to the ones obtained with the Runge-Kutta method of third order but requires one additional costly search per time step.

steps	EE	AB2	AB3	HEUN	RK3	RK4	RKF2B
3	2.65E-1	failure	failure	5.73E-2	3.22E-2	3.05E-2	3.57E-2
4	1.88E-1	1.25E-1	1.46E-1	4.17E-2	3.08E-2	3.01E-2	3.11E-2
5	1.53E-1	8.68E-2	7.75E-2	3.57E-2	3.04E-2	3.01E-2	3.04E-2
7	1.15E-1	5.67E-2	4.31E-2	3.20E-2	3.02E-2	3.01E-2	3.01E-2
10	8.64E-2	4.16E-2	3.30E-2	3.08E-2	3.01E-2	3.01E-2	3.01E-2
20	5.42E-2	3.23E-2	3.02E-2	3.02E-2	3.01E-2	3.01E-2	3.01E-2
30	4.42E-2	3.10E-2	3.01E-2	3.01E-2	3.01E-2	3.01E-2	3.01E-2
40	3.97E-2	3.06E-2	3.01E-2	3.01E-2	3.01E-2	3.01E-2	3.01E-2
50	3.72E-2	3.04E-2	3.01E-2	3.01E-2	3.01E-2	3.01E-2	3.01E-2
70	3.46E-2	3.02E-2	3.01E-2	3.01E-2	3.01E-2	3.01E-2	3.01E-2
100	3.30E-2	3.01E-2	3.01E-2	3.01E-2	3.01E-2	3.01E-2	3.01E-2

steps	EE	AB2	AB3	HEUN	RK3	RK4	RKF2B
3	2.77E0	failure	failure	2.12E-1	1.65E-1	1.61E-1	1.86E-1
4	7.18E-1	6.23E-1	9.23E-1	1.74E-1	1.63E-1	1.61E-1	1.64E-1
5	4.09E-1	3.10E-1	3.41E-1	1.69E-1	1.62E-1	1.61E-1	1.62E-1
7	3.05E-1	2.38E-1	2.09E-1	1.65E-1	1.61E-1	1.60E-1	1.61E-1
10	2.62E-1	1.92E-1	1.66E-1	1.63E-1	1.61E-1	1.60E-1	1.61E-1
20	2.11E-1	1.67E-1	1.61E-1	1.61E-1	1.60E-1	1.60E-1	1.60E-1
30	1.94E-1	1.63E-1	1.61E-1	1.61E-1	1.60E-1	1.60E-1	1.60E-1
40	1.86E-1	1.62E-1	1.60E-1	1.61E-1	1.60E-1	1.60E-1	1.60E-1
50	1.81E-1	1.61E-1	1.60E-1	1.60E-1	1.60E-1	1.60E-1	1.60E-1
70	1.75E-1	1.61E-1	1.60E-1	1.60E-1	1.60E-1	1.60E-1	1.60E-1
100	1.70E-1	1.61E-1	1.60E-1	1.60E-1	1.60E-1	1.60E-1	1.60E-1

Table 3.1.1: Quality measures Q_0 (upper table) and Q_∞ (lower table) for different step sizes and IVP solvers for test problem 2.3.1, 4,096 elements

steps	EE	AB2	AB3	HEUN	RK3	RK4	RKF2B
3	3.02E-1	failure	failure	5.70E-2	1.92 E-2	1.14E-2	4.18E-2
4	2.02E-1	1.38E-1	1.61E-1	3.27E-2	1.20E-2	6.94E-2	2.10E-2
5	1.58E-1	8.62E-2	8.45E-2	2.10E-2	1.02E-2	5.52E-3	1.32E-2
7	1.14E-1	4.74E-2	3.93E-2	1.16E-2	6.41E-3	4.19E-3	6.87E-3
10	8.10E-2	2.52E-2	1.77E-2	6.51E-3	3.95E-3	3.65E-3	4.07E-3
20	4.22E-2	8.16E-3	4.75E-3	3.90E-3	3.56E-3	3.55E-3	3.57E-3
30	2.88E-2	5.18E-3	3.69E-3	3.70E-3	3.55E-3	3.55E-3	3.56E-3
40	2.21E-2	4.31E-3	3.57E-3	3.61E-3	3.55E-3	3.55E-3	3.55E-3
50	1.80E-2	3.98E-3	3.56E-3	3.59E-3	3.55E-3	3.55E-3	3.55E-3
70	1.33E-2	3.74E-3	3.55E-3	3.57E-3	3.55E-3	3.55E-3	3.55E-3
100	9.84E-3	3.63E-3	3.55E-3	3.56E-3	3.55E-3	3.55E-3	3.55E-3

steps	EE	AB2	AB3	HEUN	RK3	RK4	RKF2B
3	2.93E0	failure	failure	2.47E-1	1.98E-1	9.71E-2	4.76E-1
4	9.77E-1	7.58E-1	1.03 E0	1.79E-1	1.16E-1	5.67E-2	1.91 E-1
5	5.77E-1	4.17E-1	4.67E-1	1.35E-1	7.95E-2	4.14E-2	1.02 E-1
7	3.10E-1	2.35E-1	2.54E-1	8.39E-2	4.67E-2	4.13E-2	5.42 E-2
10	1.86E-1	1.20E-2	1.40E-1	5.27E-2	4.15E-2	4.13E-2	4.22 E-2
20	1.16E-1	5.51E-2	4.45E-2	4.23E-2	4.13E-2	4.13E-2	4.14 E-2
30	9.18E-2	4.78E-2	4.22E-2	4.17E-2	4.13E-2	4.13E-2	4.13 E-2
40	7.94E-2	4.50E-2	4.17E-2	4.15E-2	4.13E-2	4.13E-2	4.13 E-2
50	7.19E-2	4.37E-2	4.15E-2	4.14E-2	4.13E-2	4.13E-2	4.13 E-2
70	6.32E-2	4.26E-2	4.14E-2	4.14E-2	4.13E-2	4.13E-2	4.13 E-2
100	5.70E-2	4.19E-2	4.13E-2	4.13E-2	4.13E-2	4.13E-2	4.13 E-2

Table 3.1.2: Quality measures Q_0 (upper table) and Q_∞ (lower table) for different step sizes and IVP solvers for test problem 2.3.1, 65,536 elements

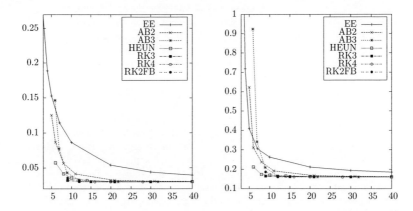

Figure 3.1.6: Quality measures Q_0 (left) and Q_∞ (right) vs. number of vector field evaluations per node for test problem 2.3.1, $4,096$ elements

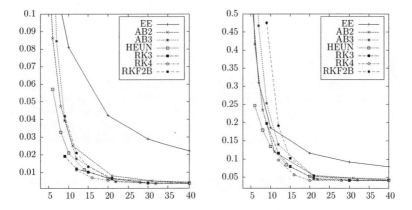

Figure 3.1.7: Quality measures Q_0 (left) and Q_∞ (right) vs. number of vector field evaluations per node for test problem 2.3.1, $65,536$ elements

The right hand side of the ODE in the deformation method is formed by a piecewise rational function which is continuous but not differentiable in a classical sense. Therefore, the right hand side of the ODE lacks the regularity needed for higher order IVP solvers to establish full convergence. Note, that for Runge-Kutta methods, the solution u must fulfil $u \in C^{k+1}$ in order to guarantee convergence of order k. The same holds for a linear k-step method. Now, we investigate if and to what extend the convergence order of the high order methods suffers decay. To do so, we compute test problem 2.3.1 with fixed step size $1 \cdot 10^{-6}$ employing the classical RK3 scheme on a tensor product grid. The deviation ρ from this reference grid induced by the IVP solver is measured by taking the l^{∞}-norm of the vector containing the pointwise deviation defined by $||x_{\text{ref}} - x||$ where x_{ref} stands for the position of the corresponding point on the reference grid. Note that our reference grid is *not* the grid obtained by applying the deformation algorithm without any approximation, as the deformation vector field is computed as recovered gradient of an FEM solution as before. However, in the deformation algorithm, the error stemming from the PDE approximation is fully decoupled from the error coming from the IVP approximation. This allows to investigate these errors separately and thus justifies the numerical tests presented here. Figures 3.1.8 and 3.1.9 which visualize ρ vs. the number of evaluations of the ODE right hand side indicate that all high order IVP methods are of second order only in the example considered. However, although being of second order only, the high order methods (AB3, RK3, RK4) perform best in the sense that given a fixed number of right hand side evaluations, the absolute approximation error related to these methods is smaller than the one of low order methods. The RKF2B method seems to be comparable to the high order Runge-Kutta methods in this numerical test. Note that when considering ρ, the advantages of the Runge-Kutta methods in favour of the linear multi-step methods are far less pronounced than the quality measures indicated before. Remarkably, the Runge-Kutta method RK4 produces less accurate results per evaluation of the right hand side than the (actually lower order) RK3-method for this example.

The results presented so far indicate that of all the IVP solvers investigated, the Runge-Kutta method of third order (RK3) is best suited for an application in the deformation method. Albeit being of second order only, it turns out to be robust and sufficiently accurate. Therefore, we will restrict ourselves to this method only.

Remark 3.1.11. *All ODE solvers considered here are explicit methods. We do not take implicit methods in account, because the successful computation of the IVPs with our explicit methods indicates that these IVPs do not exhibit stiff behaviour. Therefore, there is no justification to bear the increased computational effort connected with implicit ODE solvers.*

Besides of the error induced by the approximate solution of the IVPs, the accuracy of grid deformation is affected by the error coming from replacing the exact deformation vector field v by its discrete and therefore inexact counterpart

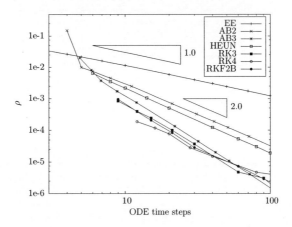

Figure 3.1.8: Deviation ρ and order of convergence of several ODE methods in the case of test problem 2.3.1, 4,096 elements

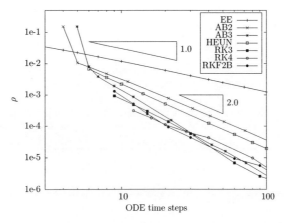

Figure 3.1.9: Deviation ρ and order of convergence of several ODE methods in the case of test problem 2.3.1, 65,536 elements

	SPR	PPR	INT
Q_0	$3.05 \cdot 10^{-2}$	$2.89 \cdot 10^{-2}$	$2.89 \cdot 10^{-2}$
Q_∞	$1.68 \cdot 10^{-1}$	$1.66 \cdot 10^{-1}$	$1.66 \cdot 10^{-1}$
Q_0	$3.60 \cdot 10^{-3}$	$3.41 \cdot 10^{-3}$	$3.41 \cdot 10^{-3}$
Q_∞	$4.43 \cdot 10^{-2}$	$4.41 \cdot 10^{-2}$	$4.41 \cdot 10^{-2}$

Table 3.1.3: Quality measures for test problem 2.3.1 for different gradient reconstruction schemes, 4,096 elements (top) and 65,536 elements (bottom)

v_h (compare eq. (2.27)). In our implementation, we stick to solving the Poisson equation, employing conforming bilinear Finite Elements and taking its gradient (raw or postprocessed) as approximation of v. Therefore, choosing an appropriate method of gradient recovery or gradient reconstruction plays a crucial role in our deformation method. Thus, we will now investigate the effect of different gradient recovery techniques to the robustness and accuracy of our deformation method. In the following, we consider three different methods of gradient reconstruction:

INT The reconstructed gradient is defined as conforming bilinear FE-function. Its values in the grid nodes are defined as the arithmetic mean of the raw gradients on the elements adjacent to this grid point.

SPR The gradient is reconstructed using the SPR-method (for details, see the introduction, section 1.1).

PPR The gradient is reconstructed using the PPR-method (for further explanations, we refer to the introduction).

In order to assess the quality of these aforementioned techniques, we again compute test problem 2.3.1 as a benchmark with 1000 equidistant RK3-steps. The results from above assure that thus there is no significant contribution of IVP solve to the overall error. The corresponding quality measures are presented in table 3.1.3. It turns out that surprisingly the quality measures of the grids are very similar which indicates that all methods perform equally well in terms of grid quality. Therefore, it is not possible to decide which method of gradient reconstruction is best suited for grid deformation based on the numerical test presented.

After these tests of components for the numerical realisation of the deformation algorithm, we return to investigations on the overall convergence behaviour. For ease of implementation, we will replace from now on in the computations of the quality measures the analytical monitor function f by its bilinear interpolant f_h on the deformed grid which is scaled then in order to fulfil equation (3.2). This leads to the scaled interpolant $\hat{f}$. Note that the integrals of f and f_h do not need to coincide which requires the additional scaling. The following lemma justifies our approach as it is shown that the additional error due to this replacement is of higher order.

Lemma 3.1.12. *Let the assumptions in theorem 3.1.10 be fulfilled. Let Q denote the quality measures computed with the analytical monitor function f and the same measure $\hat{Q}$ obtained using $\hat{f}$ instead. Here, Q stands for either Q_0 or Q_∞. Then,*

$$Q = \hat{Q} + \mathcal{O}(h^2).$$

Proof. Because of $f \in \mathcal{C}^1(\Omega)$, we know

$$\max_{x \in \Omega} |f(x) - f_h(x)| \leq Ch^2.$$

Let us denote the factor the function f_h is scaled with by ω. Then,

$$\omega \int_\Omega f_h(x)\, dx = |\Omega| = \int_\Omega f(x)\, dx,$$

and thus

$$
\begin{aligned}
\omega &= \frac{\int_\Omega (f - f_h)(x)\, dx}{\int_\Omega f(x)\, dx} + 1 \\
&\leq \frac{C|\Omega|}{\int_\Omega f_h(x)\, dx} h^2 + 1
\end{aligned}
$$

and vice versa

$$
\begin{aligned}
\frac{1}{\omega} &= \frac{\int_\Omega (f_h - f)(x)\, dx}{\int_\Omega f(x)\, dx} + \frac{\int_\Omega f(x)\, dx}{\int_\Omega f(x)\, dx} \\
&\leq Ch^2 + 1.
\end{aligned}
$$

Therefore, $\exists\, 0 < c_1, C_1 : 1 - c_1 h^2 \leq \omega \leq 1 + C_1 h^2$. Thus, we conclude $\omega - 1 = \mathcal{O}(h^2)$ and because of this

$$
\begin{aligned}
\max_{x \in \Omega} |f(x) - \hat{f}(x)| &= \max_{x \in \Omega} |f - (1 + \mathcal{O}(h^2))f_h(x)| \\
&\leq \max_{x \in \Omega} |f(x) - f_h(x)| + \mathcal{O}(h^2) \max_{x \in \Omega} |f_h(x)|.
\end{aligned}
$$

Consequently, we end up with

$$Q = \left\| \frac{f}{g} - 1 \right\| \leq \left\| \frac{(f - \hat{f})}{g} \right\| + \left\| \frac{\hat{f}}{g} - 1 \right\| = \hat{Q} + \mathcal{O}(h^2).$$

$\square$

With a similar argument, one can show that the replacement of f by $\hat{f}$ in the grid deformation itself does not alter the deformation method substantially as well. As thus the change from f to $\hat{f}$ does not have any significant influence on the convergence and the quality measures, we from now on implicitly assume this replacement for all numerical examples and experiments, writing f instead of $\hat{f}$ then.

Remark 3.1.13. *In practical computations, the computation of Q_0 requires numerical integration which introduces an additional error. Note that even on tensor product meshes the quotient of the bilinear functions f and* area *is not bilinear but rational. Because of this, it is impossible with standard cubature rules to avoid any cubature error unlike in the case of bilinear functions. In our examples, we employ the tensor product Simpson rule for computing Q_0 which is of third order. Thus, the cubature error can be asymptotically neglected. To compute Q_∞, we use the cubature points of the Simpson rule as sample points.*

Corollary 3.1.14. *Let us assume that our numerical method for computing the deformation IVPs is of second order as indicated by the numerical tests presented above. Let all initial grids be size- and shape-regular and fulfil the similarity condition (3.5) For the monitor function f and the sequence of deformed grids, the assumptions of theorem 3.1.10 may hold. Furthermore, let us choose the step size Δt of the deformation IVPs as $\Delta t = \mathcal{O}(h)$. Then, if $\|v - v_h\|_\infty = \mathcal{O}(h^2)$ and if the IVP method is of second order at least,*

$$Q_0 \leq ch, \quad Q_\infty \leq ch. \tag{3.17}$$

Proof. Due to the choice of IVP time steps, we have (in the notation of theorem 3.1.10), $\|X_h - \tilde{X}\| = \mathcal{O}(h^2)$ for all grid vertices. Thus, the preliminaries of theorem 3.1.10 are fulfilled with $\delta = 1$. □

Note that the convergence rate of our deformation algorithm with respect to Q_∞ is bounded by one due to the inevitable consistency error emerging from approximating the piecewise constant area distribution by a continuous function (compare theorem 3.1.10). Thus, from the point of view of convergence, it neither makes sense to compute the deformation vector field v with a higher order than two nor one benefits from increasing the IVP time steps stronger than $\mathcal{O}(h^{-1})$, provided the IVP solver is of second order. Moreover, there is no improvement by employing IVP methods with convergence order greater than two, as because of the roughness of v, this potential convergence order cannot be observed in practical computations. Therefore, corollary 3.1.14 refers to the *"least demanding deformation algorithm exhibiting optimal convergence"*. However, these considerations aim at the convergence order only and do not take into account e.g. the size of the corresponding error constants which lead us to take the RK3 IVP solver, which is of third order actually.

Remark 3.1.15. *On equidistant tensor product grids, all three mentioned methods of gradient recovery are known to be of second order. However, this is proven for the L^2-norm only, but not necessarily for the stronger L^∞-norm and may be even wrong for this norm. Unfortunately, we need convergence of second order in the L^∞-norm to apply theorem 3.1.10. This gap could be closed employing biquadratic finite elements for the deformation PDE, which gradient is known to be* *of second order. However, using higher order FEM for the auxiliary problem "grid deformation" contradicts our intention of combining computational efficiency with grid adaptation.*

To verify our convergence results, we compute test problem 2.3.1 using INT to obtain the deformation vector field v_h and compute the deformation IVPs with RK3. The sequence of initial grids consists of successively refined tensor product grids which are composed of four macros. Note that in this test setting, there is no consistency error for the representation of the area distribution of the initial mesh in contrast to the general case. On the coarsest grid with $NEL = 256$, we perform 3 RK3-steps, on the grid with $NEL = 1{,}024$, we employ 5. From this level of refinement on, the number of IVP steps doubles per refinement. In figure 3.1.10, we display the quality measures Q_0 and Q_∞ and their rate of decay β_0 and β_∞ respectively, depending on the number of elements NEL. Clearly, we observe $Q_\infty = \mathcal{O}(h)$ as expected even if the assumptions of theorem 3.1.10 are not valid. Note that the monitor function in our test problem is Lipschitz-continuous, but not differentiable. For Q_0, we even experience $Q_0 = \mathcal{O}(h^{3/2})$ which indicates that theorem 3.1.10 is not optimal with respect to this particular quality measure. The reason for this behaviour is subject of ongoing research. Moreover, the lower diagram in figure 3.1.10 demonstrates that the sequence of deformed grids is edge-length regular as stated in our theorem. Additionally, due to the order of $h_{\min} := \min_{e \in \mathcal{E}} |e|$, we can conclude that $|e| \geq ch \ \forall e \in \mathcal{E}$. Here, the symmetry condition (3.9) is not fulfilled which was necessary in lemma 3.1.8 to prove this property. This indicates that the symmetry condition is necessary due to technical reasons only.

Figure 3.1.10 allows the comparison between the quality measures Q obtained by calculating with the true analytical monitor function f and $\hat{Q}$ employing the bilinear counterpart $\hat{f}$. There is no significant difference between these numbers which confirms the assertions of lemma 3.1.12.

Our numerical investigation exhibited that the number of IVP time steps needs to increase as $h^{-1} = \mathcal{O}(\sqrt{N})$, if the sequence of deformations is to converge with first order. Because of this, we have to perform overall $N^{3/2}$ IVP time steps which makes the complexity of our grid deformation method grow superlinearly with respect to N. However, the computational times of the naive approach using a fixed number of time steps grows superlinearly as well (compare section 2.3). This is due to the increasing search path length which makes a single search of a point in the grid more demanding on fine grids than on coarse ones. Let us now compare these two approaches with respect to complexity. Again, we assume the sequence of initial grids to fulfil the similarity condition. On a fixed grid contained in this sequence, we denote the naive approach to approximate the deformation mapping Φ with a fixed number of time steps k by Φ_n, whereas the more sophisticated approximation to Φ utilising $K(h) = \mathcal{O}(h^{-1})$ time steps is referred to as Φ_s. Both mappings represent numerical deformations. Let us for an arbitrary grid point X denote by $\Phi_n^{(l)}(X)$ the position of this point after

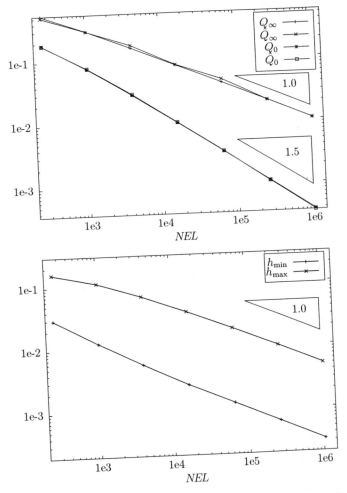

Figure 3.1.10: Quality measure Q_0 and Q_∞ computed with the true monitor function f and quality measures $\hat{Q}_0, \hat{Q}_\infty$ using the bilinear interpolant $\hat{f}$ vs. number of elements for test problem 2.3.1 (top); $h_{\min}$ and $h_{\max}$ vs. number of elements (bottom)

the l-th time step, i.e. $\Phi_n^{(0)}(X) = X = \Phi_s^{(0)}$ and $\Phi_s(X) = \Phi_s^{(K(h))}(X)$. We now assume that $|X - \Phi_s(X)| \leq C|X - \Phi(X)| \ \forall X$ with C not depending on h. This reflects the similarity of the numerical (but convergent) and the exact deformation. Furthermore, let

$$|\Phi_s^{(l)}(X) - \Phi_s^{(s+1)}| = \mathcal{O}(1/K(h)) \quad \forall \, s \in 0, \ldots K(h) \tag{3.18}$$

hold, i.e. the movement of the grid points is asymptotically not concentrated in one single time step. As the initial grid is size regular and edge-length regular, both the connecting lines $\overline{X\Phi(X)}$ and $\overline{X\Phi_s(X)}$ intersect with $\mathcal{O}(\sqrt{N}) = \mathcal{O}(h^{-1})$ elements. Thus, we con conclude from assumption (3.18) that

$$
\begin{aligned}
|\Phi_s^{(r)}(X) - \Phi_s^{(r+1)}(X)| &= \frac{\mathcal{O}(|X - \Phi_s(X)|)}{K(h)} = \frac{\mathcal{O}(|X - \Phi(X)|)}{K(h)} \\
&= \mathcal{O}(\sqrt{N})\mathcal{O}(h^{-1}) = \mathcal{O}(1)
\end{aligned}
$$

and thus, we can expect the average amount per search to be bounded in $N = \mathcal{O}(1/h)$. Note that in section 2.3, we revealed the direct connection of the search path length and the overall complexity in raytracing and distance searching. As $\mathcal{O}(N^{3/2})$ searching steps are carried out during the deformation process, the total complexity of searching during deformation is $\mathcal{O}(N^{3/2})$ for both Φ_n and Φ_s. As the PDE solve can be performed with $\mathcal{O}(N)$, *both Φ_n and Φ_s are of the same asymptotic complexity* given that our assumptions are valid. However, the sequence of deformations associated with Φ_s converges while the one connected with Φ_n does not. We validate our considerations for the well-known test problem 2.3.1 for which the convergence of the numerical deformations related to Φ_s has been shown by numerical experiments. Employing the same parameter settings as before, we apply 3 RK3-steps on the coarsest grid consisting of 256 elements and 5 RK3 steps on the grid with 1,024 elements. For every subsequent refinement, we double the number of IVP time steps. As anticipated, the average search path length for these settings is bounded (figure 3.1.11). Therefore, for this test problem, we establish first order convergent grid deformation with complexity $N^{3/2}$. This complexity is well-known from Krylov-space methods like e.g. the method of conjugated gradients. In contrast to this, multigrid methods are capable to solve the discretised PDEs with optimal complexity $\mathcal{O}(N)$ exploiting grid hierarchy. This raises the question whether such acceleration can be achieved in grid deformation as well if the sequence of initial grids was created by subsequent regular refinement. The positive answer of this question is given later in section 3.5.

3.2 Accuracy improvements by correction iterations

The convergence analysis of our grid deformation method we established in the last section provides an insight into the role the single error sources play in de-

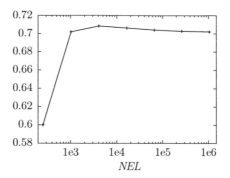

Figure 3.1.11: Average search path length for test problem 2.3.1, number of time steps increasing with grid refinement

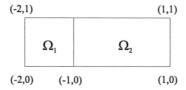

Figure 3.2.1: Regular refinement of this coarse grid produces a sequence of meshes with the discontinuous area function g defined in (3.19).

formation. However, although this analysis is consistent in itself, there are some weaknesses with respect to practical computations.

In contrast to the monitor function, there are no regularity requirements on the function g defined by the similarity condition (3.5) stated in theorem 3.1.10 besides being in $L^2(\Omega)$ and strictly positive. For general coarse grids, this function g does not need to be continuous at all. This demonstrates a sequence of grids created by regular refinement of the coarse grid shown in figure 3.2.1. The function g is (up to a constant) uniquely defined by

$$g(x) = \begin{cases} 1 & , \quad x \in \Omega_1 \\ 2 & , \quad x \in \Omega_2 \end{cases} . \tag{3.19}$$

In this situation, its scaled reciprocal and thus the right hand side of the deformation PDE (2.8) is discontinuous on a line inside the domain and thus $\notin H^1(\Omega)$. This does not contradict the statement in section 2.3 that on a given grid, the numerical area distribution function computed by bilinear interpolation of the cell sizes is Lipschitz-continuous even if g itself is discontinuous. These Lipschitz-constants are not bounded in h.

Therefore, the right hand side *lacks the regularity necessary for approximating the*

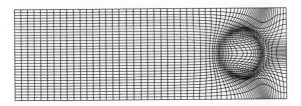

Figure 3.2.2: Resulting grid for test problem 3.2.1, 2,048 elements

vector field v on the initial mesh with convergence order greater than one required in our theorem. As a remedy, it is possible to create an auxiliary mesh for computing the deformation vector field with the desired accuracy. However, this requires meshes locally refined by either h-adaptivity with hanging nodes or by grid deformation. The former we want to avoid as it destroys the tensor product structure with its favourable properties, the latter one would have to be done manually, as our grid deformation method is likely not to converge in this situation.

Without using an auxiliary mesh, the assumption $\|v - v_h\|_\infty = \mathcal{O}(h^2)$ in corollary 3.1.14 which states the convergence for our realisation of the deformation method, does not hold. Here, v_h is defined as postprocessed gradient of an FEM solution computed with conforming bilinear elements. However, on arbitrary meshes, we cannot expect our methods of gradient recovery to be of second order, even if the solution of the deformation PDE is sufficiently smooth. This fact is known for the L^2-norm and holds particularly for the stronger L^∞-norm we consider in the deformation context. The computation of the following test problem confirms our scepticism.

Test Problem 3.2.1. *We consider the tensor product grid depicted in figure 3.2.1 which consists of two macros. As monitor function, we choose the same monitor function as in test problem 2.3.1,*

$$f(x) = \min\left\{1, \max\left\{\frac{|d - 0.25|}{0.25}, \varepsilon\right\}\right\}, \quad d := \sqrt{(x_1 - 0.5)^2 + (x_2 - 0.5)^2}.$$

The parameter ε is set to 0.1. This setting implies that on the deformed grid the largest cell has 10 times the area of the smallest one. In figure 3.2.2, we show a resulting grid consisting of 2,048 elements.

We compute this test example with the same settings used in our previous computations regarding convergence. The corresponding diagrams in figure 3.2.3 reveal that the quality measures do not converge to 0 in contrast to test problem 2.3.1. However, for the length of the element edges, it obviously still holds $c\sqrt{N} \leq h \leq C\sqrt{N}$, and thus the sequence of deformed grids remains edge-regular.

The investigation of the convergence behaviour of our deformation method lets us gain insight in the accuracy requirements for the components in the deformation algorithm, but for practical computations, convergence with respect to the

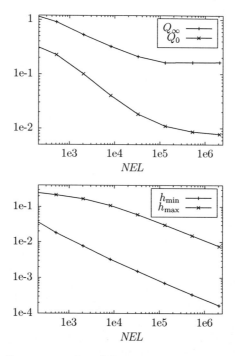

Figure 3.2.3: Quality measures Q_0 and Q_∞ vs. number of elements for test problem 3.2.1 (top); h_{min} and h_{max} vs. number of elements (bottom)

cell size h does not play a central role. Usually, there is no sequence of grids, but one grid only. Our convergence theory does not permit any prediction of the numerical value of the quality measures for a given grid; moreover, the asymptotic statements above do not give any hint how to choose the parameters associated with grid deformation, e.g. the number of IVP time steps in a certain situation. Thus, it is much more important to fulfil the condition $Q < TOL$ on the given grid regardless of any convergence. Here, TOL describes an user-defined parameter.

Therefore, we proceed a different way from now on. In a first step, we perform the grid deformation algorithm 2.3.5 computing the deformation PDE (2.8) using conforming bilinear elements and one of the proposed methods of gradient recovery. The initial value problems are then solved with one of the IVP methods introduced using a fixed number of time steps prescribed by the user. If the grid created by applying the basic grid deformation algorithm 2.3.5 does not meet the quality criterion prescribed by the user, i.e. $Q > TOL$, there are different ways to improve the grid quality. Here and in the following, Q symbolises either Q_0 or Q_∞. For instance, it is possible to increase the number of IVP time steps or to compute a better approximation to the deformation vector field v. This can be done employing higher order Finite Elements or using a further refined mesh. We will not go in this direction in this thesis, as by such measures, the numerical amount would grow considerably. Doing so, the auxiliary problem "grid deformation" would need a significant part of the overall computational time which is contradicting to its intended role as simple and cheap auxiliary method. Notice that the order of convergence cannot be improved in general by these methods, but the quality measures can be lowered. Therefore, we proceed another way.

To further improve the cell size adjustment, we repeat the deformation algorithm 2.3.5 on the newly deformed grid until $Q < TOL$ is reached. We refer to this kind of iterated deformation as *correction iteration*. The possibility of defining such an algorithm is one of the biggest advantages of our method over Liao's method, which requires the starting grid to have elements with equal area and therefore prevents any iteration process. Unfortunately, even with iterating the deformation process, it is by principle reasons impossible to achieve $Q = 0$ exactly, which is caused by the fact that in reality the area distribution is a piecewise constant function. Therefore, the interpolation error in computing g remains and prevents Q from converging to 0. To reflect this fact in the numerical algorithm, we prescribe the maximal number of correction steps N_c.

Algorithm 3.2.2 (accurate grid deformation).

input:
- f: *monitor function*
- $GRID$: *computational grid*
- N_c: *maximal number of correction steps*
- TOL: *tolerance for Q*

output:
- $GRID$: *deformed grid*

NEL	Q_0	β_0	Q_∞	β_∞
256	$1.19 \cdot 10^{-1}$	-	$3.03 \cdot 10^{-1}$	-
1,024	$3.47 \cdot 10^{-2}$	3.43	$1.31 \cdot 10^{-1}$	2.31
4,096	$9.36 \cdot 10^{-3}$	3.70	$5.60 \cdot 10^{-2}$	2.33
16,384	$2.47 \cdot 10^{-3}$	3.79	$2.14 \cdot 10^{-2}$	2.62
65,536	$6.60 \cdot 10^{-4}$	3.74	$7.92 \cdot 10^{-3}$	2.70
262,144	$2.08 \cdot 10^{-4}$	3.18	$4.10 \cdot 10^{-3}$	1.93

Table 3.2.1: Quality measures Q_0 with their factors of decay β for test problem 2.3.1, 30 correction steps applied, for $NEL = 65,536$ and $NEL = 262,144$: 50 correction steps applied

function AccurateDeformation(f, $GRID$, N_c, TOL) : $GRID$

> DO $i = 1, N_c$
>> $GRID :=$ **Deformation**(f, $GRID$)
>> $Q := Q(GRID)$
>> IF $(Q < TOL)$ EXIT LOOP
> ENDDO
> RETURN $GRID$

END AccurateDeformation

Example 3.2.3. *To assess the expected improvements in accuracy by applying the deformation algorithm 3.2.2, we again consider test problem 2.3.1. Figures 3.2.4 and 3.2.5 depict the quality parameter Q_0 and Q_∞, respectively, vs. the number of deformation cycles for different levels of refinement (NEL = number of elements). It becomes evident that the error caused by the approximate computation of the transformation Φ can be eliminated by iterating, but the deviation of Q from 0 caused by the interpolation of the area function remains. Similar to the computations in example 2.3.2, we chose RK3 as IVP-solver with time step size $1/10$, the deformation vector field v was obtained employing the SPR-technique.*

Note that in the initial step in example 3.2.3, the quality measures on the grid consisting of 262,144 elements are not smaller than the quality measures on the grid with 65,536 elements in contrast to the behaviour on coarser levels. As the gradient recovery is superconvergent on the uniform initial grid, the error induced by solving the deformation PDE decreases by a factor of four per regular refinement and thus cannot cause the observed behaviour. Very likely, the main contribution to the numerical error comes from the IVP solver, where 10 Runge-Kutta-3 steps were prescribed regardless of the level of refinement. Therefore, the IVP-induced error remains approximately constant with respect to grid refinement and thus dominates the PDE error at a certain point. However, our algorithm 3.2.2 balances the unfavourable choice of IVP steps in this situation by iterating the basic deformation algorithm. This is indicated by the quality numbers after

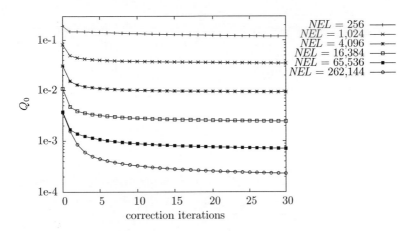

Figure 3.2.4: Quality measure Q_0 vs. number of correction iterations for test problem 2.3.1

some iterations which are significantly below the ones for the coarser grid.

Therefore, the user *needs not to care about the exact time step size to choose any more when aiming at a certain error tolerance* in deformation, as the numerical error induced by this choice vanishes by the iteration process in algorithm 3.2.2. However, for very big step sizes it may still happen that the numerical error caused by the IVP solver is so large that the resulting grid becomes invalid.

From table 3.2.1 in which the quality measure Q_0 for test problem 2.3.1 after 30 correction steps is depicted, one can conclude that the convergence of the quality measures observed in the section 3.1 can be regained by our correction iteration. Again, we observe that $Q_0 = \mathcal{O}(h^{3/2})$ and $Q_\infty = \mathcal{O}(h)$. As for the fine grids with 65,000 elements and more, 30 correction iterations are not sufficient to reach an asymptotic grid, we have performed 50 correction steps in this case.

In practical computations, it turns out to be sufficient to perform at most two correction steps. In the case of test problem 2.3.1, when computed on a tensor product grid with 1,024 elements and using 10 RK3-steps, there is visually merely a small difference between the grid obtained without correction and the one after one single correction step; the grids created using one and two correction steps do not exhibit any visual difference at all (see figure 3.2.6). In contrast to this, the quality measure Q_0 decays from $8.02 \cdot 10^{-2}$ (without correction) to $4.62 \cdot 10^{-2}$ (first correction step) and finally $4.13 \cdot 10^{-2}$ (second correction step).

Remark 3.2.4. *Figures 3.2.4 and 3.2.5 reveal that the finer the grid is, the more correction steps are needed to reach an asymptotic state, i.e. to have a converged mesh. This reminds of the iterative solution of linear systems arising from FEM for second order elliptic problems, where the number of iteration (= correction)*

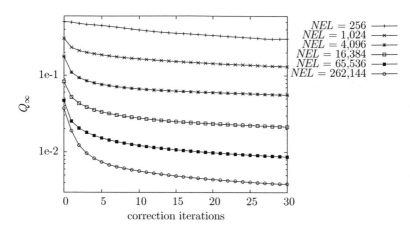

Figure 3.2.5: Quality measure Q_∞ vs. number of correction iterations for test problem 2.3.1

steps grows with mesh refinement. In this field, this unfortunate phenomenon can be overcome by applying multigrid methods which are known to be of optimal complexity, as the number of iterative steps is bounded with respect to the grid refinement level. The similarity of the two phenomena suggests to apply certain multigrid techniques to grid deformation, too. This approach will be investigated in detail in section 3.5.

Remark 3.2.5. *The accuracy improvement by correction iterations is not for free with respect to the computational time. As every correction step means applying the full basic deformation algorithm 2.3.5, the computational effort grows significantly compared to the basic grid deformation algorithm. Assuming a constant effort per basic grid deformation call, applying k correction steps may enlarge the computational by a factor of $k + 1$. Moreover, reaching an asymptotic state will lead to superlinear growth of the runtime with grid refinement, because the necessary number of correction steps grows with the refinement level (compare remark 3.2.4). We will revisit the investigation of accuracy aspects in section 3.5 for the multilevel deformation described there.*

3.3   Enhancing robustness

When applying very harsh deformation, e.g. in the case of monitor functions with extremely steep gradients or monitor functions implying extreme variations in element size, it may happen that the corresponding vector field v cannot be resolved

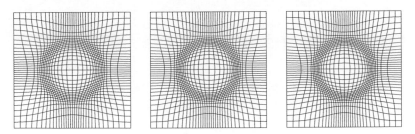

Figure 3.2.6: Computational grid after zero, one and two correction steps (test problem 2.3.1, 1,024 elements)

properly on the undeformed given grid. Thereby, the numerical grid deformation process can be disturbed to that extent that the deformed elements are not convex any more. This is not due to theoretical limitations of the method, but due to the numerical error induced by the approximate solution of the Poisson problem (2.27). In this context, the error produced by the numerical solution of the IVP (2.28) seems to be far less critical. To investigate the robustness of our deformation method with respect to desired sharp concentrations of grid cells in one region of the mesh, we investigate again test problem 2.3.1. The computations are carried out on tensor product meshes of various levels of refinement, the deformation PDE is solved using conforming bilinear Finite Elements. As IVP solver, we choose the RK3 method and a constant time step size of 0.02. Decreasing the regularisation parameter ε in the monitor function (2.29) leads to an increasing concentration of the mesh around the circle. The results in table 3.3.1 show the lowest regularisation parameter ε_{min} for which the deformation process leads to valid grids with respect to the grid size as well as to the method for gradient recovery in the deformation algorithm. To determine ε_{min}, we compute test problem 2.3.1 repeatedly and decrease ε by 0.001 in every run. No correction steps were applied in these calculations. The results exhibit that regardless of the fineness of the grid to deform, it is impossible to perform grid deformation in the current test situation if ε falls below approximately 0.01. For coarser grids with 256 elements or even less, the monitor function cannot be resolved on the initial mesh at all which enables the deformation even for extremely small ε. This however results in grids which do not feature the desired area distribution. Therefore we do not take these very coarse grids into account for this test. The minimal regularisation parameters collected in table 3.3.1 do not show any strong dependency on the grid level although on increasing high levels, ε_{min} seems to decrease. The robustness of the deformation method does not seem to be affected much by the choice of the method for gradient recovery, at least not for the given test example. In figure 3.3.1, the resulting grid with $NEL = 1{,}024$ for $\varepsilon = 1.8 \cdot 10^{-2}$ is displayed. The elements around the circle are nearly non-convex, the maximal angle measured in

NEL	INT	SPR	PPR
1,024	$1.8 \cdot 10^{-2}$	$5.0 \cdot 10^{-3}$	$1.8 \cdot 10^{-2}$
4,096	$1.9 \cdot 10^{-2}$	$2.0 \cdot 10^{-2}$	$1.9 \cdot 10^{-2}$
16,384	$2.0 \cdot 10^{-2}$	$1.7 \cdot 10^{-2}$	$2.0 \cdot 10^{-2}$
65,536	$1.5 \cdot 10^{-2}$	$1.5 \cdot 10^{-2}$	$1.5 \cdot 10^{-2}$
262,144	$1.2 \cdot 10^{-2}$	$1.2 \cdot 10^{-2}$	$1.2 \cdot 10^{-2}$
1,048,576	$9.0 \cdot 10^{-3}$	$9.0 \cdot 10^{-3}$	$9.0 \cdot 10^{-3}$

Table 3.3.1: Minimal regularisation parameters $\varepsilon_{\min}$ for which test problem 2.3.1 can be computed

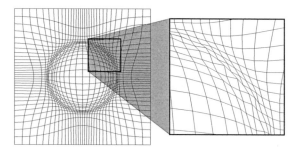

Figure 3.3.1: Resulting grid for test problem 2.3.1, $\varepsilon = 1.8 \cdot 10^{-2}$, 1,024 elements.

this grid is 179.89°.

Repeating this numerical test with a smaller constant step size of 0.002 yields exactly the same results and thus the numerical error induced by the IVP solver does not cause the failure of the deformation method. Because of this, the main numerical error in the deformation method must be introduced by the approximate solution of the deformation PDE (2.27).

As a remedy, one can solve the Poisson problem (2.27) in the deformation algorithm in a more accurate manner, e.g. by using higher order Finite Elements or a highly refined mesh. These approaches require demanding calculations and are therefore incompetetive. To overcome the limitation described here, we iterate algorithm **AccurateDeformation** using the *blended monitor function* f_s defined by

$$f_s(x) := sf(x) + (1 - s)g(x), \quad s \in [0, 1] \tag{3.20}$$

instead of the monitor function itself such that the composition of all these *adaptation steps* yields the desired deformed grid. For compatibility reasons, we require f and g to have the same integral mean value here. The following algorithm precisely defines this extension of our deformation method.

Algorithm 3.3.1 (Grid deformation with combined correction).

input: • N_a: *number of adaptation steps*
           • *GRID: computational grid*
           • $N_c(i)$: *maximal number of correction steps in i-th adaptation step*
           • *TOL(i) : tolerance for Q in i-th adaptation step*
           • *S(i): blending parameter in i-th adaptation step, $S(N_a) = 1$*

output: • *GRID: deformed grid*

function RobustDeformation(f**, *GRID*, N_c, N_a, *TOL*, *S*) : *GRID***

    `DO` $i = 1, N_a$

        GRID := **AccurateDeformation(**$f_{S(i)}$**, *GRID*, $N_c(i)$, *TOL(i)***)

    `ENDDO`

    `RETURN` *GRID*

END RobustDeformation

Here, the question arises how to choose the *blending parameter s* in formula (3.20) and how many adaptation steps to perform. Looking at the monitor function f for the well-known test problem 2.3.1 defined in equation (2.29), it turns out that the parameter ε, which describes the ratio of the area of the smallest and largest elements on the deformed grid is at the same time a certain measure for the difficulty of the deformation: For $\varepsilon = 1$, the deformation is trivial as $\Phi = Id$ because of $f \equiv 1$, the choice $\varepsilon = 0.1$ served as first test example; for $\varepsilon = 0.01$, the deformation method in its original form is likely to fail as investigated above, and deforming a grid according to f with $\varepsilon = 0$ would lead to elements with area zero and is therefore impossible for principle reasons. Let now $f_{\min}$ and $f_{\max}$ denote the maximum and the minimum of the arbitrary monitor function f in Ω. With $\varepsilon \approx f_{\min}/f_{\max}$ for the monitor function (2.29), it is reasonable to regard the ratio $f_{\max}/f_{\min}$ as indicator for the numerical difficulty of the grid deformation. These considerations, however, are valid when starting from an initial mesh with constant area distribution only. As example, we do not expect any difficulties with the deformation method when tackling the test problem 2.3.1 with small ε on a suitably predeformed grid, where the actual deformation is almost trivial. Note that in such a situation it holds $f(x) \approx g(x)$ regardless of the properties of both f and g. This gives the hint that the deviation of $f(x)/g(x)$ from a constant value might be linked to the amount of numerical difficulties of the deformation problem. This is confirmed by rearranging the transformation equation (2.1) to

$$|J\Phi(x)| = \frac{f(\Phi(x))}{g(x)},$$

which suggests to define

$$\gamma' := \frac{\max_{x \in \Omega} \frac{f(\Phi(x))}{g(x)}}{\min_{x \in \Omega} \frac{f(\Phi(x))}{g(x)}} \geq 1$$

as heuristical difficulty measure. Unfortunately, γ' is unsuitable for practical computations as it relies on the unknown transformation Φ. Therefore, neglecting the influence of Φ, we define

$$\gamma := \frac{(f(x)/g(x))_{\max}}{(f(x)/g(x))_{\min}} \geq 1$$

as indicator for the numerical problems to expect during deformation. For the trivial deformation problem $f = g$, it holds $\gamma = \gamma' = 1$. Note that in the special case of test problem 2.3.1, we have $g = 1$ after scaling, and therefore it then holds $\gamma = f_{\max}/f_{\min} \approx \varepsilon^{-1}$, and the initial "difficulty measure" is recovered.

Remark 3.3.2. *It is likely that for certain situations, the influence of Φ cannot be neglected, and that γ is considerably smaller than γ'. To obtain an upper bound for γ', it is straightforward to define*

$$\hat{\gamma} := \frac{\frac{\max_{x\in\Omega} f(x)}{\min_{x\in\Omega} g(x)}}{\frac{\min_{x\in\Omega} f(x)}{\max_{x\in\Omega} g(x)}} = \frac{\max_{x\in\Omega} f(x) \cdot \max_{x\in\Omega} g(x)}{\min_{x\in\Omega} f(x) \cdot \min_{x\in\Omega} g(x)} \geq \gamma.$$

However, in the case $f = g$ i.e. in the case of a trivial deformation process, we end up with

$$\hat{\gamma} = \frac{\max_{x\in\Omega} f^2(x)}{\min_{x\in\Omega} f^2(x)}$$

which is overly pessimistic and does not reflect that the deformation is actually trivial.

With these considerations in mind, it is reasonable to use the blending of formula (3.20) in order to *reduce γ to such an extent that every partial deformation according to the blended monitor function in algorithm 3.3.1 can be handled by deformation algorithm 3.2.2.*
We now assume that algorithm 3.2.2 is able to produce valid grids for all settings with $1 < \gamma < \gamma_0$ as indicated by the findings presented in table 3.3.1. Further on, we take for granted that for the given monitor function $\bar{f}$ and the given current area distribution $\bar{g}$, it holds $\bar{\gamma} > \gamma_0$. Let $0 < \alpha \leq 1$ be a safety factor. In this situation, N_a is set to

$$N_a = \left\lceil \frac{\ln(\bar{\gamma})}{\ln\alpha + \ln(\gamma_0)} \right\rceil, \tag{3.21}$$

where $\lceil x \rceil := \min_{k\in\mathbb{Z}}\{k \geq x\}$ denotes the ceiling function. The blending parameter s_i to define the blended monitor function (compare formula (3.20)) in the i-th grid adaptation step is chosen such that

$$\gamma_i = \left(\sqrt[N_a]{\bar{\gamma}} \right)^i \leq (\alpha\gamma_0)^i \tag{3.22}$$

holds. Note that γ_i here is with respect to the initial area distribution $\bar{g}$. Starting on the undeformed mesh, we perform a sequence of N_a deformation steps where each of these steps aims to fulfil

$$g_i(x)|\Phi_i(x)| = f_{s_i}(\Phi_i(x)), \quad i = 1,\dots N_a. \tag{3.23}$$

Because of

$$g_i(x) = f_{s_{i-1}}(\Phi_{i-1}(x)), \tag{3.24}$$

we can conclude that for the composition transformation $\Phi := \Phi_{N_a} \circ \ldots \circ \Phi_1$ which implicitly has been applied to the initial grid, it holds

$$
\begin{aligned}
|J\Phi(x)| &= |J(\Phi_{N_a} \circ \ldots \circ \Phi_1(x))| \\
&= |J(\Phi_{N_a})(\Phi_{N_a-1} \circ \ldots \circ \Phi_1(x)) \cdot J(\Phi_{N_a-1} \circ \ldots \circ \Phi_1(x))| \\
&= |J(\Phi_{N_a})(\Phi_{N_a-1} \circ \ldots \circ \Phi_1(x))| \ldots |J\Phi_1(x)| \\
&\overset{(3.23)}{=} \frac{f_{s_{N_a}}(\Phi_{N_a} \circ \ldots \circ \Phi_1(x))}{g(\Phi_{N_a-1} \circ \ldots \circ \Phi_1(x))} \frac{f_{s_{N_a-1}}(\Phi_{N_a-1} \circ \ldots \circ \Phi_1(x))}{g(\Phi_{N_a-2} \circ \ldots \circ \Phi_1(x))} \ldots \frac{f_{s_1}(\Phi_1(x))}{g(x)}
\end{aligned}
$$

which leads, if $s_{N_a} = 1$, to the deformation equation (2.1). Notice, that by virtue of equations (3.22) and (3.23) as well as (3.24), every single deformation step is computable with algorithm 3.2.2. In the derivation above, we assume that equation (3.23) holds exactly for the transformations Φ_i computed, whereas in fact these equations are fulfilled approximately only. But, due to the internal correction steps of the grid deformation algorithm 3.2.2, this error can be controlled by applying sufficiently many correction steps.

Lemma 3.3.3. *The blending parameter s_i in the i-th adaptive step has to be chosen as*

$$s_i = \frac{\left(\sqrt[N_e]{\overline{\gamma}} \right)^i - 1}{\left(\frac{f}{g} - 1 \right)_{\max} - \gamma_i \left(\frac{f}{g} - 1 \right)_{\min}}. \tag{3.25}$$

In the case of an equidistributed initial grid, the computation of s_i simplifies to

$$s_i = \frac{\left(\sqrt[N_e]{\overline{\gamma}} \right)^i - 1}{f_{\max} + \left(\sqrt[N_e]{\overline{\gamma}} \right)^i (1 - f_{\min}) - 1}. \tag{3.26}$$

Proof. Starting from

$$\frac{(f_{s_i}/g)_{\max}}{(f_{s_i}/g)_{\min}} \overset{!}{=} \gamma_i,$$

we obtain

$$
\begin{aligned}
\gamma_i &= \frac{(f_{s_i}/g)_{\max}}{(f_{s_i}/g)_{\min}} \\
&= \frac{\left(s_i \frac{f}{g} + \frac{(1-s_i)g}{g} \right)_{\max}}{\left(s_i \frac{f}{g} + \frac{(1-s_i)g}{g} \right)_{\min}} \\
&= \frac{s_i \left(\frac{f}{g} - 1 \right)_{\max} + 1}{s_i \left(\frac{f}{g} - 1 \right)_{\min} + 1},
\end{aligned}
$$

from which by virtue of equation (3.22) the statement (3.25) follows. For an equidistributed initial mesh, we can assume that $g \equiv 1$ due to the scaling of f and

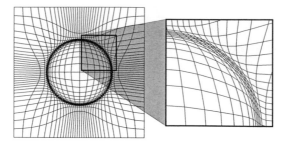

Figure 3.3.2: Computational grid for test problem 2.3.1, $\varepsilon = 0.01$, 1,024 elements

g to the same integral value. Thus, statement (3.26) is an immediate consequence of equation (3.25). □

Remark 3.3.4. *For the trivial deformation with $f = g$, the formulas (3.25) and (3.26) are not defined. This reflects the fact that it does not make sense to define several adaptation steps in this case. Furthermore, for the trivial deformation it holds $N_a = 1$ because of $\gamma = 1$ and formula (3.21). As we require that $s_{N_a} = 1$, there is no need to compute any blending parameter in this situation.*

Remark 3.3.5. *Utilising the improved algorithm 3.3.1, it is possible to compute the well-known test problem 2.3.1 for $\varepsilon = 0.01$ without any difficulties (see e.g. figure 3.3.2). Here, we prescribe $\gamma_i = 10$ resulting in 2 adaptive steps and compute the test problem on a tensor product grid with 1,024 elements. In contrast to this, algorithm 3.2.2 leads to invalid grids in this case regardless of the IVP solver, the number of time steps and the method for gradient recovery.*

The choice of γ_0 is of largely heuristic nature and depends on many factors. It should be made clear, that it is a heuristic approach to measure the "amount of difficulty" in a single number γ only and that therefore there is no mathematically rigorous way to compute γ_0. The numerical tests performed at the beginning of this section lead to the practical guess of $\gamma_0 = 10$ which is also chosen for all subsequent test problems featuring several adaptation steps in this thesis. The safety factor α we set to 1.0 in all our computations. Although without rigorous theoretical justification, for practical computations, the break-up into several less harsh deformation problems guided by the heuristical difficulty measure γ and the choice of adaptation steps described above leads to a significant gain of robustness and thus enlarges the class of solvable deformation problems considerably.

However, computing the well-known test problem 2.3.1 with $\varepsilon = 0.001$ still leads to tangled grids in spite of the improvements of the deformation algorithm introduced in this section. In the subsequent section, we will investigate this phenomenon more in detail connected with the presentation of approaches to overcome the failure of the grid deformation in this situation.

ε	$\alpha_{\min}$	$\alpha_{\max}$	asp. rat.	Q_0	Q_∞
$1.0 \cdot 10^{-1}$	36.77	144.54	8.11	$1.44 \cdot 10^{-2}$	$1.66 \cdot 10^{-1}$
$7.0 \cdot 10^{-2}$	29.94	151.26	9.55	$1.66 \cdot 10^{-2}$	$1.42 \cdot 10^{-1}$
$5.0 \cdot 10^{-2}$	25.13	156.49	11.07	$2.00 \cdot 10^{-2}$	$1.60 \cdot 10^{-1}$
$2.0 \cdot 10^{-2}$	12.77	171.13	15.76	$3.14 \cdot 10^{-2}$	$2.56 \cdot 10^{-1}$
$1.0 \cdot 10^{-2}$	7.42	176.44	19.45	$4.23 \cdot 10^{-2}$	$4.17 \cdot 10^{-1}$
$7.0 \cdot 10^{-3}$	5.02	178.97	22.64	$4.64 \cdot 10^{-2}$	$2.94 \cdot 10^{-1}$
$5.0 \cdot 10^{-3}$	3.64	179.46	24.73	$5.26 \cdot 10^{-2}$	$3.89 \cdot 10^{-1}$
$2.0 \cdot 10^{-3}$	-	-	-	-	-
$1.0 \cdot 10^{-3}$	-	-	-	-	-
$7.0 \cdot 10^{-4}$	-	-	-	-	-
$5.0 \cdot 10^{-4}$	-	-	-	-	-

Table 3.4.1: Quality measures Q_0 and Q_∞ as well as minimal and maximal angles $\alpha_{\min}$ and $\alpha_{\max}$ and aspect ratios for test problem 2.3.1 with varying ε ("$-$" indicates that a test leads to invalid grids).

3.4 Postprocessing and further stabilisation

In the preceding sections, we developed the improved grid deformation method 3.3.1, which is capable to provide grids of (nearly) arbitrary area distribution with controllable accuracy. However, computing test problem 2.3.1 with regularisation parameter $\varepsilon = 0.001$ still results in invalid grids due to non-convex elements. To understand the reason for this phenomenon, we compute this test problem for decreasing ε with grid deformation algorithm 3.3.1 on a tensor product grid with 4,096 elements. The number of adaptation steps is chosen according to formula (3.21), the blending parameters according to equation (3.26), the deformation vector field is recovered using INT. After the actual deformation, one correction step is applied. As in the preceding computations, we apply 10 RK3-steps with constant stepsize 0.1 as IVP solver.

In table 3.4.1, we present the quality measures Q_0 and Q_∞ as well as the minimal ($\alpha_{\min}$) and maximal angle ($\alpha_{\max}$) of the deformed grid and the maximal aspect ratio. The small quality measures reveal that the resulting grids feature the desired area distribution regardless of ε, but with decreasing ε, some elements inside the grid become more and more detoriated up to an extent where reasonable FE calculations with parametric Finite Elements are not possible any more. In our test example, the resulting grids contain non-convex elements for $\varepsilon < 5.0 \cdot 10^{-3}$. This is indicated by "$-$" in table 3.4.1. This process is indicated by large maximal angles approaching $180°$ with decreasing ε and small minimal angles tending to 0. Obviously, the observed failure is connected with the fact that our grid deformation method does not provide so far any possibility to control the *quality* of the produced grid itself in terms of prescribing limits for e.g. the maximal aspect ratio or maximal angle in an element. Unfortunately, this is one principal limitation of our approach which aims at the element area only by prescribing $|J\Phi|$. Instead, controlling features like e.g. the eigenvalues or eigenvectors of $J\Phi$ would lead to control over element shape and element alignment in the grid. Notwithstanding there are deformation methods which aim at controlling these quantities described in the literature (we refer e.g. to [23] and the references cited therein), we will proceed another way, as all these methods are numerically much more demanding than our deformation method.

We enhance our deformation method 3.3.1 by applying suitable *postprocessing* after every deformation substep in order to improve the grid quality if necessary. If the postprocessing affects the element area distribution, this will be indicated by large quality measures and will thus be balanced by the correction steps automatically applied to reach the tolerance for the quality measure.

Algorithm 3.4.1 (Robust grid deformation with combined correction and postprocessing).

input:
- f: monitor function
- $GRID$: computational grid
- N_a: number of adaptation steps
- $N_c(i)$: maximal number of correction steps in i-th adaptation step
- $TOL(i)$: tolerance for Q in i-th adaptation step
- $S(i)$: blending parameter in i-th adaptation step, $S(N_a) = 1$

output:
- $GRID$: deformed grid

function **ImprovedDeformation**$(f,\ GRID,\ N_a,\ N_c,\ TOL,\ S) : GRID$
    DO $i = 1, N_a$
        $GRID := $ **Deformation**$(f_{S(i)},\ GRID)$
        $GRID := $ **PostProcess**$(GRID)$
        DO $j = 1, N_c(i)$
            IF $(j > 1)$ $GRID := $ **PostProcess**$(GRID)$
            $GRID := $ **Deformation**$(f_{S(i)},\ GRID)$
            $Q := Q(GRID)$
            IF $(Q < TOL(i))$ EXIT LOOP
        ENDDO
    ENDDO
    RETURN $GRID$

END ImprovedDeformation

This motivates the need of a suitable postprocessing which improves the grid quality without affecting the area distribution. To do so, we consider optimisation-based mesh smoothing following [43]. For a given grid vertex v, we denote the set of grid vertices adjacent to v by $\mathcal{V}(v)$ and set the minimal angle to $\alpha_{\min}(v) := \min_{w,w' \in \mathcal{V}(v)}\{\angle(w, v, w')\}$. Now, we solve the local optimisation problem

$$\alpha_{\min}(v) \rightarrow \max \qquad (3.27)$$

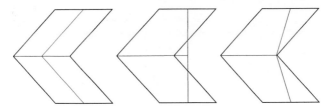

Figure 3.4.1: Invalid elements by angle "optimisation": initial grid (left), grid after moving v to $\tilde{v}$ (middle), grid after moving v to v^* (right).

with the side condition that the resulting grid must not contain tangled elements. In order to solve optimisation problem (3.27), we consider the intersection $\tilde{v}$ of the lines connecting the left/right and the lower/upper adjacent vertices. In the case of an admissible logical tensor product mesh, it holds $\tilde{v} = v^*$, if the side condition is not violated by $\tilde{v}$. Here, v^* denotes the true solution of the optimisation problem (3.27). The simple example in figure 3.4.1 shows that $\tilde{v}$ may in fact violate the side condition. Therefore, if one of the elements adjacent to v is degenerated after moving v to $\tilde{v}$, the new coordinates are rejected and v remains at the old position.

Remark 3.4.2. *For the test grid displayed in figure 3.4.1, v^* is shown in the rightmost image. Even though none of the elements is degenerated, two of the four elements have an edge of length (almost) zero which renders the result of true optimisation process as incompetitive as taking $\tilde{v}$. Because of this, we do not solve the optimisation problem (3.27) in an exact manner at all but restrict ourselves to the much faster computable "candidate" $\tilde{v}$.*

Remark 3.4.3. *The grids displayed in figure 3.4.1 moreover show that the angle-based optimisation we employ can lead to non-convex elements on non-convex domains. This shortcoming can be overcome by optimising the grid according to a modified functional, which incorporates the minimal angle but additionally penalises small or even negative lengths of element edges. However, the domain of the test problem we consider is the convex unit square and therefore we stick to the grid optimisation method described above.*

In the context of grid deformation, the postprocessing is employed in order to avoid "almost invalid" grids like the ones considered above. Therefore, it seems favorable to apply the postprocessing only in regions of the grid with bad mesh quality, i.e. with very small or very large angles. Because of this we modify the postprocessing based on angle optimisation. We compute a blending parameter $t_{\min}$ being a linear function in $\alpha_{\min}$ such that $t_{\min}(0) = 1$ and $t_{\min}(\pi/2) = 0$. In an analogous way, we determine the blending parameter $t_{\max}$. Then, the new position of v is obtained by

$$v_{\text{new}} := (1-t)v + t\tilde{v}, \quad t := \max\{t_{\min}, t_{\max}, 0\}.$$

Figure 3.4.2: Optimal tensor product grid in terms of minimal angle

To optimise the whole grid, we loop in a global optimisation step over all grid vertices and apply the local optimisation in the way described above. Hereby, the initial coordinates of a moved vertex are stored, such that all adjacent vertices are at their initial coordinates during the local optimisation. In contrast to common grid smoothing procedures like e.g. Laplacian smoothing (see cf. [42]) which tend to equilibrate the cell size, the optimisation based grid smoothing does not affect the area distribution which is very favourable in an algorithm which controls cell sizes. An explanation for this behaviour is given in figure 3.4.2. As the minimal angle is 90° and therefore optimal, mesh optimisation does not change the grid at all, while repeated Laplacian smoothing will produce an equidistant grid. Freitag and her coworkers showed that the combination of Laplacian smoothing and grid optimisation produces more favourable results than each method used alone [42]. Following them, we consider in what follows a combination of Laplacian grid smoothing and grid optimisation.

In order to test whether applying the postprocessing described above improves the grid quality and thus the robustness of the deformation, we repeat the computation of test problem 2.3.1 with the same parameters as before, but we apply after every adaptation step 15 Laplacian smoothing steps and 30 grid optimisation steps. After the correction step in the actual deformation method, no postprocessing takes place. The correction step compensates the influence of postprocessing to the area distribution which is affected mainly by the Laplacian part of our postprocessing. However, numerical tests revealed that using angle-based optimisation as sole postprocessing behaves inferior with respect to robustness than the combined approach. This finding is in accordance to Freitag [42]. The results of the calculations are displayed in table 3.4.2, the grid is shown in figure 3.4.3. It turns out that both the maximal angles are decreased and the minimal angles are increased resulting in the expected gain of robustness in the deformation algorithm. On the other hand, the quality measures are increased by postsmoothing. This documents the adverse effect of the postprocessing to the area distribution. However, despite the considerable gain of robustness, the tendency to produce degenerated grids cannot fully cured by our post-smoothing, as α_{min} still decreases and α_{max} still increases with decreasing ε.

To apply deformation to extremely concentrate a grid around a circle in a unit

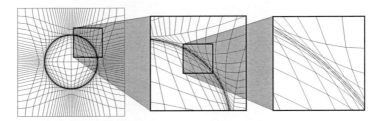

Figure 3.4.3: Resulting grid for test problem 2.3.1, $\varepsilon = 0.001$, postsmoothing with 15 Laplacian smoothing steps and 30 grid optimisation steps (grid size for display purposes reduced to 1,024 elements)

ε	$\alpha_{\min}$	$\alpha_{\max}$	asp. rat.	Q_0	Q_∞
$1.0 \cdot 10^{-1}$	40.06	140.83	6.93	$2.19 \cdot 10^{-2}$	$4.03 \cdot 10^{-1}$
$7.0 \cdot 10^{-2}$	33.59	147.45	7.80	$2.54 \cdot 10^{-2}$	$4.56 \cdot 10^{-1}$
$5.0 \cdot 10^{-2}$	28.17	152.89	8.86	$2.92 \cdot 10^{-2}$	$5.14 \cdot 10^{-1}$
$2.0 \cdot 10^{-2}$	17.17	164.13	11.87	$4.09 \cdot 10^{-2}$	$6.74 \cdot 10^{-1}$
$1.0 \cdot 10^{-2}$	11.09	170.89	14.47	$5.10 \cdot 10^{-2}$	$7.80 \cdot 10^{-1}$
$7.0 \cdot 10^{-3}$	9.39	172.69	22.20	$5.57 \cdot 10^{-2}$	$8.53 \cdot 10^{-1}$
$5.0 \cdot 10^{-3}$	7.86	174.00	30.32	$6.03 \cdot 10^{-2}$	$9.91 \cdot 10^{-1}$
$2.0 \cdot 10^{-3}$	4.83	178.07	55.64	$7.03 \cdot 10^{-2}$	$1.51 \cdot 10^{0}$
$1.0 \cdot 10^{-3}$	3.65	179.60	76.26	$7.33 \cdot 10^{-2}$	$1.99 \cdot 10^{0}$
$7.0 \cdot 10^{-4}$	4.18	178.74	70.53	$7.44 \cdot 10^{-2}$	$2.60 \cdot 10^{0}$
$5.0 \cdot 10^{-4}$	4.16	178.98	63.77	$7.60 \cdot 10^{-2}$	$2.80 \cdot 10^{0}$

Table 3.4.2: Quality measures Q_0 and Q_∞ as well as minimal and maximal angles $\alpha_{\min}$ and $\alpha_{\max}$ and aspect ratio for test problem 2.3.1 with varying ε, postsmoothing with 15 Laplacian smoothing steps and 30 steps of grid optimisation

ε	$\alpha_{\min}$	$\alpha_{\max}$	asp. rat.	Q_0	Q_∞
$1.0 \cdot 10^{-1}$	37.02	144.01	5.68	$4.23 \cdot 10^{-3}$	$1.00 \cdot 10^{-1}$
$7.0 \cdot 10^{-2}$	29.60	150.97	6.96	$4.24 \cdot 10^{-3}$	$8.57 \cdot 10^{-2}$
$5.0 \cdot 10^{-2}$	23.40	157.17	8.50	$4.90 \cdot 10^{-3}$	$1.13 \cdot 10^{-1}$
$2.0 \cdot 10^{-2}$	11.33	169.26	13.88	$7.10 \cdot 10^{-3}$	$2.40 \cdot 10^{-1}$
$1.0 \cdot 10^{-2}$	6.13	174.45	19.20	$9.32 \cdot 10^{-3}$	$3.52 \cdot 10^{-1}$
$7.0 \cdot 10^{-3}$	4.72	175.37	21.18	$9.90 \cdot 10^{-3}$	$3.23 \cdot 10^{-1}$
$5.0 \cdot 10^{-3}$	3.35	176.75	23.64	$1.11 \cdot 10^{-2}$	$3.64 \cdot 10^{-1}$
$2.0 \cdot 10^{-3}$	1.34	179.24	34.54	$1.51 \cdot 10^{-2}$	$5.62 \cdot 10^{-1}$
$1.0 \cdot 10^{-3}$	-	-	-	-	-
$7.0 \cdot 10^{-4}$	0.15	180.00	98.72	$1.98 \cdot 10^{-2}$	$5.46 \cdot 10^{-1}$
$5.0 \cdot 10^{-4}$	-	-	-	-	-

Table 3.4.3: Results of test problem 3.4.4, $NEL = 4,096$

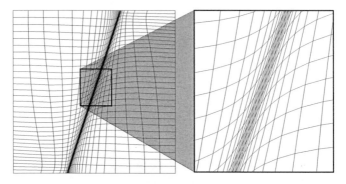

Figure 3.4.4: Resulting grid for test problem 3.4.4, $\varepsilon = 0.01$

square seems to be an rather academic test problem far away from any practical computation. In contrast, this test problem anticipates some of the problems to expect when applying grid deformation within a CFD simulation. In these simulations, often shock fronts occur which are usually unaligned to the grid. For accuracy reasons, it is essential in these computations to resolve the shocks properly which requires strong grid refinement nearby the shock fronts. This is the background of the following test problem.

Test Problem 3.4.4. *An equidistant tensor product mesh on the unit square* $[0, 1]^2$ *shall be deformed according to the monitor function*

$$f(x) = \min\{1, \max\{d, \varepsilon\}\}, \quad d := |3x - 1|.$$

ε	α_{min}	α_{max}	asp. rat.	Q_0	Q_∞
$1.0 \cdot 10^{-1}$	39.11	140.88	5.18	$6.80 \cdot 10^{-3}$	$2.16 \cdot 10^{-1}$
$7.0 \cdot 10^{-2}$	31.69	148.31	6.37	$7.73 \cdot 10^{-3}$	$2.48 \cdot 10^{-1}$
$5.0 \cdot 10^{-2}$	25.52	154.48	7.98	$8.76 \cdot 10^{-3}$	$3.38 \cdot 10^{-1}$
$2.0 \cdot 10^{-2}$	13.31	166.69	15.37	$1.19 \cdot 10^{-2}$	$7.14 \cdot 10^{-1}$
$1.0 \cdot 10^{-2}$	7.77	172.24	25.94	$1.48 \cdot 10^{-2}$	$1.12 \cdot 10^{0}$
$7.0 \cdot 10^{-3}$	6.51	173.49	41.59	$1.63 \cdot 10^{-2}$	$1.69 \cdot 10^{0}$
$5.0 \cdot 10^{-3}$	5.05	174.95	56.31	$1.79 \cdot 10^{-2}$	$2.14 \cdot 10^{0}$
$2.0 \cdot 10^{-3}$	2.44	177.57	105.45	$2.28 \cdot 10^{-2}$	$3.40 \cdot 10^{0}$
$1.0 \cdot 10^{-3}$	1.36	178.65	102.62	$2.63 \cdot 10^{-2}$	$4.06 \cdot 10^{0}$
$7.0 \cdot 10^{-4}$	1.25	178.75	101.20	$2.77 \cdot 10^{-2}$	$8.04 \cdot 10^{0}$
$5.0 \cdot 10^{-4}$	0.95	179.05	111.91	$2.89 \cdot 10^{-2}$	$9.05 \cdot 10^{0}$

Table 3.4.4: Results of test problem 3.4.4 with Laplacian smoothing and angle-based optimisation as postprocessing for $NEL = 4,096$

All parameters in the grid deformation algorithm are set as for the computation of test problem 2.3.1 before. The aspect ratio and the minimal and maximal angle in the grid α_{min} and α_{max} respectively as well as the quality measures Q_0 and Q_∞ are presented in table 3.4.3. One of the resulting grids is displayed in figure 3.4.4. For the sake of image clarity, the number of elements was reduced to 1,024 elements for display only. It turns out that the resulting grids feature the prescribed element size distribution which is indicated by small Q_0 and Q_∞. Moreover, the grid features the desired stretched elements along the oblique shock which can be derived from the increasing maximal aspect ratios. However, in spite of obtaining valid grids for almost all ε considered, with decreasing ε, the grid quality detoriates, as α_{min} tends to $0°$ and α_{max} to $180°$. For small ε, the detoriation even makes the deformation method fail as before. When applying the first correction step, the resulting grid contains tangled elements in the region of the oblique shock. Unlike the robustness issues described in section 3.3 which stem from the inability to resolve the monitor function and the deformation vector field on the initial grid properly and which could be resolved by increasing the number of adaptation cycles, the failure here is caused by the bad mesh quality which prevents accurate gradient computation on the resulting grid. Therefore, neither increasing the number of adaptation steps nor the number of time steps in the IVP solver nor changing the gradient recovery routine makes the deformation process work as numerical experiments show, because in all cases, one ends up with (almost) the same grid which does not permit suitable FEM calculations. This makes the correction step fail. Again, the application of 10 Laplacian smoothing steps and 20 grid optimisation steps improves the robustness of the grid deformation algorithm at the price of worse quality numbers.

Although it is clearly visible that introducing postsmoothing leads to increased robustness of the grid deformation, it also becomes clear that postprocessing can extend the limitations of our deformation method, but it cannot remove them. It is a essential drawback of our method that it allows no control of grid properties like minimal or maximal angle, aspect ratio or the alignment of certain elements. However, in most of the practical cases, these limitations are not severe.

3.5 Speed-up by exploiting grid hierarchy

In the previous sections, we developed a grid deformation method which is able to produce grids of nearly arbitrary area distribution and which turned out to be robust and accurate. To be suitable for practical computations, however, besides accuracy and robustness, additionally the computational time consumed by the deformation algorithm has to be taken into account. It is clearly desirable that the deformation time is only a small fraction of the overall computation time. To achieve this, we have to test every main component of our deformation algorithm with respect to numerical complexity and time consumption.

To solve the global Poisson problem (2.8) in the deformation method, we employ multigrid solvers which are known to grow linearly with the number of unknowns and therefore are of optimal complexity. The same applies obviously for the actual IVP solve, given a constant number of IVP time steps. In contrast to this, the amount for searching the grid during the IVP solve will grow superlinearly, this operation is of complexity $\mathcal{O}(N^{3/2})$ (compare section 2.3). Therefore, the time to search the grid will dominate the overall time consumed by the deformation algorithm on fine grids.

Remark 3.5.1. *In order to obtain convergence, it is necessary to double the number of time steps per refinement step of the initial grid. Because of this, the search path length remains bounded and the asymptotic complexity is $\mathcal{O}(N^{3/2})$ like in the case of constant time step size (compare sections 3.1 and 3.2). However, the former choice of the time step size implies by far longer computational times than the latter one. Therefore, we aim now at accelerating the deformation method in the case of constant time step size and neglect in a first step the convergence behaviour.*

We again compute test problem 2.3.1 with $\varepsilon = 0.1$ on an equidistant tensor product grid consisting of four macros with deformation algorithm 3.3.1. We perform one adaptation step according to formula (3.21) and one correction step. No postprocessing is employed. As IVP solver, we apply 10 RK3-steps with equidistant step size and search the grid by raytracing search. The deformation PDE is solved with multigrid which performs four ADITRIGS-steps [76] for pre- and postsmoothing, respectively. The ADITRIGS smoother is known to be quite demanding numerically, but is superiorly robust for almost arbitrarily deformed grids. The corresponding coarse grid problems are solved with the method of conjugated gradients (CG). The multigrid iteration is stopped when the norm of the residuum falls below 10^{-9}. The tests were computed on an AMD Opteron

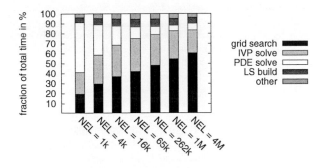

Figure 3.5.1: Relative timings for selected components of grid deformation method 3.3.1

250 server equipped with a 64-bit Linux operating system. The code is compiled with the Intel Fortran Compiler v. 9.1. with full optimisation. As predicted, the relative timings in figure 3.5.1 confirm the anticipated dominance of the time for raytracing search caused by its superlinear growth. Thus, searching the grid indicated by "grid search" in the diagram takes roughly 65% of the overall computational time on the finest grid, whereas the actual IVP solve ("IVP solve") and the assembly of the linear system for the Poisson equation ("LS build") form a small part only. The same holds for solve time of the deformation Poisson equation ("PDE solve"). All other operations in grid deformation like scaling the monitor function, computing the current area distribution or computing quality measures are embraced in "other".

Remark 3.5.2. *In this particular example, our multigrid method grows even sub-linear on coarse levels. This suggests in figure 3.5.1 that the actual search time as well as the assembly time for the linear system would grow superlinearly which is not the case. They all show linear growth as expected.*

Accelerating the deformation method therefore implies decreasing the overall search time. So far, we did not exploit in our examples that the sequence of initial grids is constructed by repeated regular refinement of one coarse grid, which is a much stronger condition than the similarity condition (3.5) considered so far. The most straightforward way to utilise grid hierarchy is to employ hierarchical searching. As pointed out before, however, we render hierachical searching methods unsuitable for our purposes because of their inherent overhead.

We exploit the given mesh hierarchy in another way. Let us assume that our initial grid resulted from refining the coarse one k times regularly, and let us denote this initial grid by $\mathcal{T}_k$. We refer to the grid created by applying grid deformation to $\mathcal{T}_k$ as $\tilde{\mathcal{T}}_k$. In contrast to previous variants of grid deformation, we aim now at deforming $\mathcal{T}_{k-1}$ instead of $\mathcal{T}_k$ itself. After one regular refinement of $\tilde{\mathcal{T}}_{k-1}$, the resulting grid $\hat{\mathcal{T}}_k$ will resemble $\tilde{\mathcal{T}}_k$ closely for sufficiently large k. This indicates that on $\hat{\mathcal{T}}_k$, all grid points are near the position they take on $\tilde{\mathcal{T}}_k$. Note that compared to the deformation algorithms presented so far, the average search path length is cut in half in this approach. Applying another deformation step to $\hat{\mathcal{T}}_k$ to improve the grid quality will not lead to significant changes of the node positions and therefore we can expect short average search paths on level k resulting in short search times on the finest grid. Like in section 2.3, we define the search path length as the number of element traversals during the search, and the average search path length as fraction of the total number of searched points and the total number of element traversals in all searches. Boundary points and intermediate points of the Runge-Kutta-method are neglected. Iterating this two-level approach leads to *multilevel deformation* which we present in algorithmic form.

Algorithm 3.5.3 (multilevel deformation).

input: • f: *monitor function*
  • $i_{\min}$: *coarsest level for deformation*
  • $i_{\max}$: *compute level*
  • i_{incr}: *level increment*
  • $N_a(i)$: *number of adaptation steps on refinement level i*
  • $N_{\mathrm{pre}}(i)$: *number of presmoothing steps on refinement level i*
  • $GRID$: *computational grid*
  • $N_c(i, j)$: *maximal number of correction steps for the j-th adaptation steps on multigrid level i*
  • $TOL(i, j)$: *tolerance for Q in the j-th adaptation step on refinement level i*
  • $S(i, j)$: *blending parameter in j-th adaptation step on refinement level i, $S(i, N_a) = 1$*

output: • $GRID$: *deformed grid*

function MultilevelDeformation(f, $GRID$, N_c, N_a, N_{pre}, TOL, S) : $GRID$

$\quad GRID_{i_{\min}} := \mathbf{RESTRICT}(GRID, i_{\min})$

$\quad \mathtt{DO}\ i = i_{\min}, i_{\max}, i_{\mathrm{incr}}$

$\qquad GRID_i := \mathbf{PreSmooth}(\ GRID_i, N_{\mathrm{pre}}(i))$

$\qquad GRID_i := \mathbf{ImprovedDeformation}(f, GRID_i, N_c(i, :), N_a(i, :),$
$\qquad\qquad\qquad\qquad\qquad\qquad\qquad TOL(i, :), S(i, :))$

$\qquad \mathtt{IF}\ (i < i_{\max})\, GRID_{i+1} := \mathbf{PROLONGATE}(GRID_i)$

$\quad \mathtt{ENDDO}$

$\quad GRID := GRID_{i_{\max}}$

$\quad \mathtt{RETURN}\ GRID$

END ImprovedDeformation

In the algorithm presented above, we denote by PROLONGATE and RE-STRICT regular refinement and coarsening of our computational grid, respectively. The coarsening is performed by omitting all vertices which do not belong to the fine grid. By PreSmooth, we denote a generic grid smoothing strategy, which can be realised e.g. by Laplacian smoothing or the local grid optimisation introduced in a preceding section. Both PROLONGATE, RESTRICT and PreSmooth are purely grid-related and must not be confused with the similarly named building blocks of a multigrid solution algorithm.

In what follows, we refer to the deformation algorithm 3.3.1 as "one-level deformation" in contrast to the multilevel deformation defined in algorithm 3.5.3. To evaluate the multilevel deformation method, we consider test problem 2.3.1 with

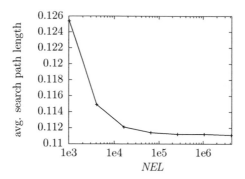

Figure 3.5.2: Average search path length vs. number of elements NEL, test problem 2.3.1 with multilevel deformation

the same settings as above, but using algorithm 3.5.3 instead. The parameters for the multilevel deformation are set as follows.

- $i_{\min} = 3$ leading to an initial grid consisting of 256 elements

- $i_{\mathrm{incr}} = 1$

- $N_a(i_{\min})$ is chosen adaptively by formula (3.21), $N_a(i) = 1 \forall i > i_{\min}$

- $N_c(i, j) \equiv 0$

- $N_{\mathrm{pre}} \equiv 2$

From figure 3.5.2, it can be deduced that in multilevel deformation, the average search path length remains bounded as desired. This is a significant improvement compared to the one-level deformation method, where the average search path length behaves like $\sqrt{N}$ (compare section 2.3, table 2.3.3). Based upon this numerical result, we now prove *optimal complexity* of the multilevel deformation.

Lemma 3.5.4. *Let us assume that in multilevel deformation, the average search path length is bounded independently of the multigrid level as indicated by our numerical tests. Let N denote the number of nodes in the grid. Then, the total number of IVP time steps N_{solve} as well as the total number N_{search} of point-in-element-tests during search grow at most like N.*

Proof. Let us denote the number of nodes on the grid on level $i_{\min}$ by N_0. Further, let us assume $i_{\mathrm{incr}} = 1$ without loss of generality. Then, the grid on the finest level consists of $N_0 \cdot 4^{(i_{\max} - i_{\min})}$ elements. As all grids on the levels in between $i_{\min}$ and $i_{\max}$ are deformed, totally

$$\sum_{i=i_{\min}}^{i_{\max}} N_0 \, 4^{i - i_{\min}}$$

grid points are moved, and thus

$$
\begin{aligned}
N_{solve} &\leq C \sum_{i=i_{\min}}^{i_{\max}} N_0 4^{i - i_{\min}} \\
&\leq C N_0 N \sum_{i=i_{\min}}^{i_{\max}} \frac{1}{N_0 4^{i_{\max} - i_{\min}}} 4^{i - i_{\min}} \\
&= C N \sum_{i=i_{\min}}^{i_{\max}} \left(\frac{1}{4}\right)^{i - i_{\max}} = C N \sum_{k=0}^{i_{\max} - i_{\min}} \left(\frac{1}{4}\right)^{k} \\
&= C N \frac{1 - \left(\frac{1}{4}\right)^{i_{\max} - i_{\min}}}{1 - \frac{1}{4}} \leq \frac{4}{3} C N,
\end{aligned}
$$

where C stands for the maximum over the number of time steps. This proves the first part of the statement. The second one follows immediately by the boundedness of the search path length. $\qquad\square$

For test problem 2.3.1, the computational time for grid search and IVP solve grows linearly (cf. figure 3.5.3) when applying multilevel deformation in contrast to $\mathcal{O}(N^{3/2})$ for the one-level deformation. Regarding the computational time for these deformation components as measure for their numerical complexity, these findings confirm the statement of lemma 3.5.4.

It turns out that on fine grids, although the multilevel deformation performs much faster than the one-level deformation method, it even produces more accurate results as indicated by smaller quality numbers (see table 3.5.4). This indicates that as assumed before, $\hat{\mathcal{T}}_k$ resembles $\tilde{\mathcal{T}}_k$ well on the intermediate grid levels such that the actual deformation on the current level plays in fact the role of a correction step on the current level. For coarse grids up to 65,000 elements, the quality measures of the multilevel deformation are inferior to the ones obtained from one-level deformation. This likely stems from the poor resolution of the desired area distribution on coarse grids. Because of this, the assumption $\hat{\mathcal{T}}_k \approx \tilde{\mathcal{T}}_k$ is "less valid" than on fine grids and therefore the deformation step on level k can only partially be regarded as correction step. However, multilevel deformation is intended for very fine grids only. These findings justify that in our example for multilevel deformation, no correction step is applied on the finest level in contrast to one-level deformation.

The multilevel deformation requires the user to choose the newly introduced parameters $i_{\min}$ and $N_{\text{pre}}(i)$. For all other free parameters, we gave rules of choice before, albeit in another context. In what follows, we investigate numerically

NEL	Q_0	Q_∞	h_{min}	α_{min}	α_{max}	runtime [s]
1,024	$5.641 \cdot 10^{-2}$	$2.241 \cdot 10^{-1}$	$1.10 \cdot 10^{-2}$	41.13	140.41	$1.74 \cdot 10^{-1}$
4,096	$2.071 \cdot 10^{-2}$	$1.146 \cdot 10^{-1}$	$5.25 \cdot 10^{-3}$	38.20	142.68	$3.97 \cdot 10^{-1}$
16,384	$7.795 \cdot 10^{-3}$	$5.732 \cdot 10^{-2}$	$2.56 \cdot 10^{-3}$	36.80	143.62	$1.07 \cdot 10^{0}$
65,536	$3.212 \cdot 10^{-3}$	$2.897 \cdot 10^{-2}$	$1.27 \cdot 10^{-3}$	36.10	144.07	$3.40 \cdot 10^{0}$
262,144	$1.421 \cdot 10^{-3}$	$1.447 \cdot 10^{-2}$	$6.32 \cdot 10^{-4}$	35.81	144.32	$1.18 \cdot 10^{1}$
1,048,576	$6.625 \cdot 10^{-4}$	$7.307 \cdot 10^{-3}$	$3.15 \cdot 10^{-4}$	35.68	144.37	$4.35 \cdot 10^{1}$
4,194,304	$3.189 \cdot 10^{-4}$	$3.680 \cdot 10^{-3}$	$1.57 \cdot 10^{-4}$	35.63	144.41	$1.73 \cdot 10^{2}$
1,024	$6.237 \cdot 10^{-2}$	$2.562 \cdot 10^{-1}$	$1.13 \cdot 10^{-2}$	44.66	136.92	$1.64 \cdot 10^{-1}$
4,096	$2.231 \cdot 10^{-2}$	$1.248 \cdot 10^{-1}$	$5.11 \cdot 10^{-3}$	39.89	140.93	$4.13 \cdot 10^{-1}$
16,384	$8.329 \cdot 10^{-3}$	$6.310 \cdot 10^{-2}$	$2.48 \cdot 10^{-3}$	38.21	142.21	$1.13 \cdot 10^{0}$
65,536	$3.386 \cdot 10^{-3}$	$3.190 \cdot 10^{-2}$	$1.23 \cdot 10^{-3}$	37.49	142.83	$3.48 \cdot 10^{0}$
262,144	$1.475 \cdot 10^{-3}$	$1.594 \cdot 10^{-2}$	$6.10 \cdot 10^{-4}$	37.20	142.98	$1.19 \cdot 10^{1}$
1,048,576	$6.807 \cdot 10^{-4}$	$8.042 \cdot 10^{-3}$	$3.05 \cdot 10^{-4}$	37.10	142.99	$4.53 \cdot 10^{1}$
4,194,304	$3.266 \cdot 10^{-4}$	$4.323 \cdot 10^{-3}$	$1.52 \cdot 10^{-4}$	37.05	142.98	$1.76 \cdot 10^{2}$
1,024	$7.957 \cdot 10^{-2}$	$2.897 \cdot 10^{-1}$	$1.23 \cdot 10^{-2}$	46.33	135.65	$9.52 \cdot 10^{-2}$
4,096	$2.355 \cdot 10^{-2}$	$1.390 \cdot 10^{-1}$	$5.28 \cdot 10^{-3}$	40.67	140.17	$3.44 \cdot 10^{-1}$
16,384	$9.030 \cdot 10^{-3}$	$7.183 \cdot 10^{-2}$	$2.53 \cdot 10^{-3}$	37.84	142.52	$1.06 \cdot 10^{0}$
65,536	$3.551 \cdot 10^{-3}$	$3.464 \cdot 10^{-2}$	$1.25 \cdot 10^{-3}$	37.00	143.29	$3.54 \cdot 10^{0}$
262,144	$1.534 \cdot 10^{-3}$	$1.735 \cdot 10^{-2}$	$6.22 \cdot 10^{-4}$	36.71	143.46	$1.24 \cdot 10^{1}$
1,048,576	$7.049 \cdot 10^{-4}$	$8.688 \cdot 10^{-3}$	$3.10 \cdot 10^{-4}$	36.61	143.47	$4.59 \cdot 10^{1}$
4,194,304	$3.372 \cdot 10^{-4}$	$4.402 \cdot 10^{-3}$	$1.55 \cdot 10^{-4}$	36.57	143.48	$1.97 \cdot 10^{2}$

Table 3.5.1: Quality measures Q_0, Q_∞, minimal element size h_{min} as well as minimal and maximal angles α_{min}, α_{max} and runtime (AMD Opteron 250), $i_{min} = 2$ (upper part), $i_{min} = 3$ (medium part) and $i_{min} = 4$ (lower part), test problem 2.3.1 with $\varepsilon = 0.1$

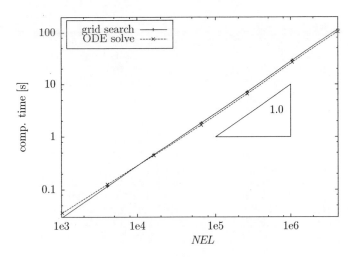

Figure 3.5.3: Computational times for grid search and IVP solve for multilevel deformation

the dependence of accuracy of the deformation process and the quality of the resulting grid from the choice of $i_{\min}$ and N_{pre}. We again compute the well-known test problem 2.3.1 with multilevel deformation 3.5.3 using the same settings as above for $i_{\min} = 2, 3$ and 4 yielding initial grids consisting of 64, 256 and 1,024 elements, respectively. The results in table 3.5.1 show that the influence of the initial refinement level is small and decreases as expected with growing grid level. This holds true both for the quality measures and the quantities measuring the grid quality. Note that regardless of the initial refinement level chosen, the quality measures are by far lower than the ones for one-level deformation (compare table 3.5.4). The difference of the run times turned out to be insignificant. These experiments indicate that the multilevel grid deformation algorithm 3.5.3 is robust with respect to the choice of $i_{\min}$ given that the deformation problem is solvable on the initial grid.

In another series of tests, we set $i_{\min}$ to 3, but vary the number of Laplacian presmoothing steps $N_{\text{pre}}(i)$. We compute the same test problem computed with the same parameter settings as above and choose the number of Laplacian presmoothing steps uniformly for all grid refinement levels to 2, 3 and 4. As the computational time for grid smoothing contributes only insignificantly to the overall runtime, we omit the runtime for these tests. The resulting grid quantities listed in table 3.5.2 permit the conclusion that the multilevel deformation algorithm is robust with respect to the choice of presmoothing steps as well. Increasing

NEL	Q_0	Q_∞	$h_{\min}$	$\alpha_{\min}$	$\alpha_{\max}$
1,024	$6.237 \cdot 10^{-2}$	$2.562 \cdot 10^{-1}$	$1.13 \cdot 10^{-2}$	44.66	136.92
4,096	$2.231 \cdot 10^{-2}$	$1.248 \cdot 10^{-1}$	$5.11 \cdot 10^{-3}$	39.89	140.93
16,384	$8.329 \cdot 10^{-3}$	$6.310 \cdot 10^{-2}$	$2.48 \cdot 10^{-3}$	38.21	142.21
65,536	$3.386 \cdot 10^{-3}$	$3.190 \cdot 10^{-2}$	$1.23 \cdot 10^{-3}$	37.49	142.83
262,144	$1.475 \cdot 10^{-3}$	$1.594 \cdot 10^{-2}$	$6.10 \cdot 10^{-4}$	37.20	142.98
1,048,576	$6.807 \cdot 10^{-4}$	$8.042 \cdot 10^{-3}$	$3.05 \cdot 10^{-4}$	37.10	142.99
4,194,304	$3.266 \cdot 10^{-4}$	$4.323 \cdot 10^{-3}$	$1.52 \cdot 10^{-4}$	37.05	142.98
1,024	$6.009 \cdot 10^{-2}$	$2.476 \cdot 10^{-1}$	$1.13 \cdot 10^{-2}$	44.43	137.21
4,096	$2.062 \cdot 10^{-2}$	$1.214 \cdot 10^{-1}$	$5.12 \cdot 10^{-3}$	39.92	140.89
16,384	$7.118 \cdot 10^{-3}$	$6.216 \cdot 10^{-2}$	$2.49 \cdot 10^{-3}$	38.27	142.18
65,536	$2.529 \cdot 10^{-3}$	$3.128 \cdot 10^{-2}$	$1.23 \cdot 10^{-3}$	37.65	142.65
262,144	$9.255 \cdot 10^{-4}$	$1.550 \cdot 10^{-2}$	$6.12 \cdot 10^{-4}$	37.40	142.77
1,048,576	$3.571 \cdot 10^{-4}$	$7.785 \cdot 10^{-3}$	$3.05 \cdot 10^{-4}$	37.29	142.80
4,194,304	$1.481 \cdot 10^{-4}$	$3.993 \cdot 10^{-3}$	$1.53 \cdot 10^{-4}$	37.23	142.81
1,024	$5.887 \cdot 10^{-2}$	$2.427 \cdot 10^{-1}$	$1.12 \cdot 10^{-2}$	44.24	137.43
4,096	$2.001 \cdot 10^{-2}$	$1.196 \cdot 10^{-1}$	$5.13 \cdot 10^{-3}$	39.86	140.94
16,384	$6.859 \cdot 10^{-3}$	$6.082 \cdot 10^{-2}$	$2.50 \cdot 10^{-3}$	38.29	142.20
65,536	$2.371 \cdot 10^{-3}$	$3.074 \cdot 10^{-2}$	$1.23 \cdot 10^{-3}$	37.69	142.60
262,144	$8.207 \cdot 10^{-4}$	$1.518 \cdot 10^{-2}$	$6.14 \cdot 10^{-4}$	37.44	142.72
1,048,576	$2.876 \cdot 10^{-4}$	$7.651 \cdot 10^{-3}$	$3.06 \cdot 10^{-4}$	37.32	142.76
4,194,304	$1.036 \cdot 10^{-4}$	$3.892 \cdot 10^{-3}$	$1.53 \cdot 10^{-4}$	37.27	142.77

Table 3.5.2: Quality measures Q_0, Q_∞, minimal element size $h_{\min}$ as well as minimal and maximal angles $\alpha_{\min}$ and $\alpha_{\max}$, $N_{\mathrm{pre}} = 2$ (upper part), $N_{\mathrm{pre}} = 3$ (medium part) and $N_{\mathrm{pre}} = 4$ (lower part), test problem 2.3.1 with $\varepsilon = 0.1$

NEL	Q_0	Q_∞	$h_{\min}$	$\alpha_{\min}$	$\alpha_{\max}$
1,024	$7.593 \cdot 10^{-2}$	$3.001 \cdot 10^{-1}$	$1.06 \cdot 10^{-2}$	43.68	137.24
4,096	$4.937 \cdot 10^{-2}$	$2.963 \cdot 10^{-1}$	$5.08 \cdot 10^{-3}$	39.40	141.72
16,384	$3.195 \cdot 10^{-2}$	$2.039 \cdot 10^{-1}$	$2.44 \cdot 10^{-3}$	37.06	143.37
65,536	$1.940 \cdot 10^{-2}$	$1.516 \cdot 10^{-1}$	$1.21 \cdot 10^{-3}$	35.66	144.73
262,144	$1.129 \cdot 10^{-2}$	$1.048 \cdot 10^{-1}$	$6.01 \cdot 10^{-4}$	35.01	145.27
1,048,576	$6.321 \cdot 10^{-3}$	$7.126 \cdot 10^{-2}$	$2.99 \cdot 10^{-4}$	34.91	145.25
4,194,304	$3.452 \cdot 10^{-3}$	$4.765 \cdot 10^{-2}$	$1.49 \cdot 10^{-4}$	34.93	145.15

Table 3.5.3: Quality measures Q_0, Q_∞, minimal element size $h_{\min}$ as well as minimal and maximal angles $\alpha_{\min}$ and $\alpha_{\max}$, no presmoothing

NEL	Q_0			Q_∞		
	one-lev	$i_{incr} = 1$	$i_{incr} = 2$	one-lev	$i_{incr} = 1$	$i_{incr} = 2$
1,024	$4.62 \cdot 10^{-2}$	$6.24 \cdot 10^{-2}$	$7.96 \cdot 10^{-2}$	$1.92 \cdot 10^{-1}$	$2.56 \cdot 10^{-1}$	$2.90 \cdot 10^{-1}$
4,096	$1.47 \cdot 10^{-2}$	$2.23 \cdot 10^{-2}$	$2.36 \cdot 10^{-2}$	$9.09 \cdot 10^{-2}$	$1.25 \cdot 10^{-1}$	$1.39 \cdot 10^{-1}$
16,384	$4.64 \cdot 10^{-3}$	$8.33 \cdot 10^{-3}$	$1.10 \cdot 10^{-2}$	$4.33 \cdot 10^{-2}$	$6.31 \cdot 10^{-2}$	$8.28 \cdot 10^{-2}$
65,536	$1.67 \cdot 10^{-3}$	$3.39 \cdot 10^{-3}$	$3.37 \cdot 10^{-3}$	$2.11 \cdot 10^{-2}$	$3.19 \cdot 10^{-2}$	$3.53 \cdot 10^{-2}$
262,144	$1.56 \cdot 10^{-3}$	$1.48 \cdot 10^{-3}$	$1.80 \cdot 10^{-3}$	$1.74 \cdot 10^{-2}$	$1.59 \cdot 10^{-2}$	$2.07 \cdot 10^{-2}$
1,048,576	$1.24 \cdot 10^{-3}$	$6.81 \cdot 10^{-4}$	$7.03 \cdot 10^{-4}$	$1.67 \cdot 10^{-2}$	$8.04 \cdot 10^{-3}$	$8.61 \cdot 10^{-3}$
4,194,304	$9.13 \cdot 10^{-4}$	$3.28 \cdot 10^{-4}$	$3.99 \cdot 10^{-4}$	$1.74 \cdot 10^{-2}$	$4.32 \cdot 10^{-3}$	$5.57 \cdot 10^{-3}$

Table 3.5.4: Quality measures for test problem 2.3.1 computed with one-level deformation 3.3.1 (left part) and multilevel deformation 3.5.3 with $i_{incr} = 1$ (medium part) and $i_{incr} = 2$ (right part)

the number of smoothing steps from 2 to 3 leads to a significant improvement of Q_0 which is particularly notable on very fine levels. However, Q_∞ changes much less, and the difference in the other grid-related quantities displayed are minor to insignificant. Further increasing the number of presmoothing to 4 has less impact than the change from 2 to 3 steps. It is even possible to set presmoothing aside at all, even though this has negative impact to the grid quality and deformation accuracy. This demonstrates the data obtained by computing our standard test problem 2.3.1 as before without any presmoothing (table 3.5.3). However, the quality measures are inferior to the ones obtained by the one-level deformation algorithm. Overall, we conclude that the multilevel deformation is robust with respect to the number of presmoothing steps, i.e. changing their number will not lead to significantly different resulting grids.

The investigations made in this chapter reveal another possibility to accelerate the grid deformation. Keeping in mind that obviously the deformation steps on the finer grids change the mesh only minimally, it seems favorable to omit some "intermediate" levels. This can be realised in algorithm 3.5.3 by enlarging the level increment. Again, we compute test problem 2.3.1 with the multilevel deformation method 3.5.3 and the parameter settings above expect of i_{incr} set to 2. For even levels of refinement, the starting level i_{min} is set to 4, for odd ones to 3 resulting in initial grids with 256 and 1042 elements, respectively. This is to guarantee that the actual level increment is always two. The quality measures displayed in table 3.5.4 show a slight decline compared to the case of $i_{incr} = 1$, but they are still much better than the ones of the one-level deformation on multigrid levels 9 and 10 (cf. table 3.5.4).

The comparison of the overall runtimes for the different deformation algorithms in figure 3.5.4 demonstrates that by multilevel deformation, a significant speed-up can be achieved. Here, we denote by OneLevConst the variant of the one-level deformation with constant size of IVP time steps independend of the refinement level, OneLevConv stands for the variant where the number of time steps doubles

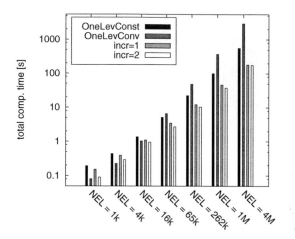

Figure 3.5.4: Runtime comparisons for one-level deformation with constant and increasing step size and multilevel deformation with level increment one and two

per refinement in order to achieve convergence. The runtimes are measured on the same computer (AMD Opteron 250) and the same settings as above. The measurements were repeated until the standard deviations fell below 1%. The results show that on high levels of refinement, OneLevConv is up to 3 times slower than OneLevConst and thus confirm the statements of remark 3.5.1. The speed-up of the multilevel deformation is the more pronounced the higher the multigrid level is, as on high levels, the search times spoil the overall efficiency by their superlinear growth for the one-level deformation.

Remark 3.5.5. *As the average search path length is bounded with respect to the grid size and moreover is very small in multilevel deformation, hierarchical grid search will not lead to any speed-up compared to "normal" search methods like the ones we presented. For one-level deformation, on sufficiently fine grids, hierarchical searching may provide a speed-up due to possibly large search path lengths, but on these grids, the multilevel deformation proved superior with respect to both accuracy and speed. So, there is ultimately no reason to consider hierarchical grid search in any of our deformation methods in order to achieve substantial speed-ups.*

Our numerical experiments revealed the superior accuracy of the multilevel deformation method regardless of the setting of its free parameters. This encourages to revisit the convergence investigations of section 3.1 and to compare the

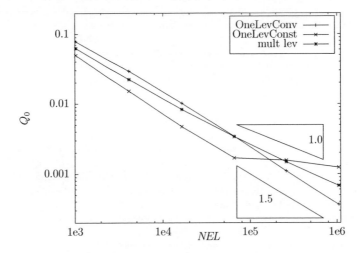

Figure 3.5.5: Q_0 vs. *NEL* for one-level and multilevel deformation methods

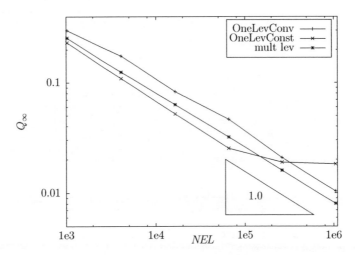

Figure 3.5.6: Q_∞ vs. *NEL* for one-level and multilevel deformation methods

results of this section with the ones from the different variants of the one-level methods considered so far. In the aforementioned section 3.1, we considered the one-level method we have called OneLevConv in the current section, where the convergence observed relies on doubling the time steps per regular refinement. In the previous section 3.2, we looked at the same method, but applied 10 ODE time steps regardless of the level (variant OneLevConst). Consequently, it did not converge at all. In this case, convergence could be obtained empirically by repeated correction steps. However, here we restrict ourselves to one correction step only. We now compare the results of these one-level methods with the results of our multi-level deformation ("mult lev") with $i_{\min} = 3$, $i_{\mathrm{incr}} = 1$ and $N_{\mathrm{pre}} = 2$. The data visualised in figures 3.5.5 and 3.5.6 clearly lead to the conclusion that *the multilevel deformation features first order convergence*, i.e. $Q = \mathcal{O}(h)$ for both Q_0 and Q_∞. Moreover, *the quality measures are comparable with OneLevConv* for all levels of grid refinement investigated. We note that the multilevel deformation does not feature the "superconvergence" of Q unlike OneLevConv, where even $Q = \mathcal{O}(h^{3/2})$ holds. The runtime comparisons in figure 3.5.4 however demonstrate that OneLevConv needs up to 15 times more computational time to achieve a similar accuracy as the multilevel approach.

As pointed out in section 3.1 before, the accuracy of any of our deformation algorithms is limited by three error sources: the consistency error, the error induced by solving the Poisson equation in the deformation algorithm numerically and the error resulting from the approximate solution of all IVPs. As our multilevel deformation does not contain any correction steps, all three errors must thus decrease with grid refinement due to the convergence observed. For the consistency error being $\mathcal{O}(h)$, this is obvious. The same holds true for the error of the solution of the deformation PDE. The somehow surprising fact that apparently the IVP-induced error decreases as well despite the constant time step size requires further investigations.

In the motivation of our multilevel method, we assumed that $\hat{\mathcal{T}}_k$ resembles $\tilde{\mathcal{T}}_k$, i.e. there is no substantial difference between deforming first and refining afterwards and vice versa. This leads us to interpret the deformation step on the intermediate level k as correction step acting on $\hat{\mathcal{T}}_k$. We can expect that the necessary corrections become smaller the finer the grid is. Let us denote the grid points of $\hat{\mathcal{T}}_k$ by x and its counterpart on $\tilde{\mathcal{T}}_k$ by X. Then, we measure the distance of these two grids by

$$d_k := \max_{x \in \mathcal{V}} ||x - X||, \quad d_k^{\mathrm{avg}} := \frac{1}{|\mathcal{V}|} \sum_{x \in \mathcal{V}} ||x - X||.$$

Here, $\mathcal{V}$ stands for the set of vertices in $\hat{\mathcal{T}}_k$. If we denote the mapping constructed by the deformation algorithm on refinement level k by Φ_k, then $\Phi_k(\hat{\mathcal{T}}_k) = \tilde{\mathcal{T}}_k$ and by construction we obtain

$$||x - \Phi_k(x)|| \leq d_k \quad \forall \, x \in \mathcal{V}.$$

Let us furthermore assume that $d_k = \mathcal{O}(h^{1+\tau})$. Like in section 3.1, we denote now the image of x by X_h, if it was obtained by solving the PDE exactly, but the corresponding IVP approximately. The image of x computed by solving both differential equations with numerical methods we refer to by $\tilde{X}$. We take now for granted that the *relative error induced by the approximate IVP solve* is bounded, i.e.

$$\frac{||X_h - \tilde{X}||}{||x - X||} \leq C \quad \forall x \in .$$

This is a by far weaker condition on the IVP solve than the true convergence we required up to now. Then, it follows $||X_h - \tilde{X}|| = \mathcal{O}(h^{1+\tau})$. Thus, the IVP-induced error decreases as $h^{1+\tau}$ *without increasing the number of IVP time steps*. Together with the decay of the other two error sources, the observation $d_k = \mathcal{O}(h^{1+\tau})$ is sufficient to explain the convergence observed. For the test problem 2.3.1, numerical experiments show that $d_k = \mathcal{O}(h^2)$ (figure 3.5.7). Far from having presented a rigorous proof, we summarize our results in following theorem.

Theorem 3.5.6. *Let $0 < \varepsilon < f \in \mathcal{C}^1(\bar{\Omega})$ be a strictly positive monitor function and $0 < g_{\min} < g < g_{\max}$. Let the sequence of initial grids $(\mathcal{T}_i)_{i \in I}$ emerge from regular refinement of one coarse grid. Then, N denoting the number of grid points, the multilevel deformation algorithm*

    a) is of optimal asymptotic complexity $\mathcal{O}(N)$

    b) converges with first order: $Q_0 = \mathcal{O}(h)$, $Q_\infty = \mathcal{O}(h)$.

To confirm that theorem 3.5.6 holds in more general situations, we go back to test problem 3.2.1. Due to the initial grid, the area function g is not continuous for the given grid which prevents the deformation method OneLevConv from converging (compare section 3.1). We now apply our multilevel deformation algorithm to this test problem setting $i_{\min} = 3$, $i_{\mathrm{incr}} = 1$ and $N_{\mathrm{pre}} = 2$. The results displayed in figure 3.5.8 clearly demonstrate the convergence of the multilevel method, more precisely, we can deduce $Q_0 = \mathcal{O}(h)$ and $Q_\infty = \mathcal{O}(h)$. This finding indicates that the multilevel algorithm is not only faster and more accurate than one-level deformation, but *more robust* as well, as it converges in a situation where the one-level deformation method does not converge. We will come back to this result in the next section which deals with applying grid deformation to the L-domain.

3.6 Towards applications: grid deformation on the L-domain

It is one of the main goals of this thesis to establish the grid deformation algorithm developed in this chapter as tool for a posteriori-driven error estimation. As preparation, we apply grid deformation to concentrate the mesh around reentrant corners. This is motivated by the following test problem.

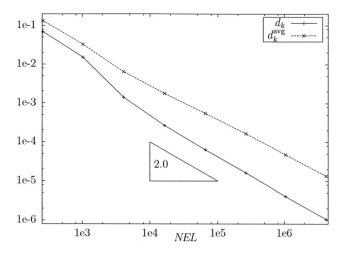

Figure 3.5.7: d_k and d_k^{avg} vs. NEL, test problem 2.3.1

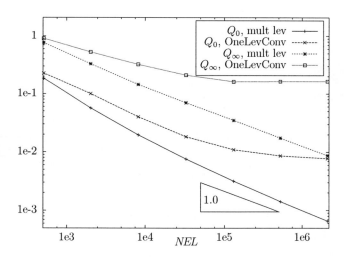

Figure 3.5.8: Comparison of the quality measures obtained by OneLevConv and multilevel deformation for test problem 3.2.1

Test Problem 3.6.1. *Consider the Laplace equation on the L-shaped domain* $\Omega := [-0.5, 0.5]^2 \setminus [0, 0.5]^2$. *The smooth Dirichlet boundary data are chosen such that in polar coordinates*

$$u(r, \varphi) = r^{2/3} \sin(\frac{2}{3}\varphi) \tag{3.28}$$

holds.

Test problem 3.6.1 is one of the most prominent problems investigated in the literature and serves as standard benchmark problem for error control and adaptive FEM. Being in $H^{1+3/4}$ only, its solution (3.28) exhibits a *corner singularity* in $(0, 0)$ and thus the gradient error is highly localised around the reentrant corner. Moreover, the lack of regularity of the solution makes the global gradient error decrease on size regular grids with $\mathcal{O}(h^{2/3}) = \mathcal{O}(\sqrt[3]{NEL})$ only in the H^1-norm instead of $\mathcal{O}(h) = \mathcal{O}(\sqrt{NEL})$ which can be expected for solutions in $H^2(\Omega)$ (for details, see [80]). The L-domain serves as prototypical example for all domains featuring reentrant corners.

As the gradient error is highly localised around $(0, 0)$, the grid points need to be concentrated around this reentrant corner in order to better resolve the singularity of the gradient compared to a size regular mesh. (we refer to [80, pp. 428] and the references cited therein). From this, the following test problem arises.

Test Problem 3.6.2. *An equidistant tensor product grid on the L-shaped domain* $\Omega := [-0.5, 0.5]^2 \setminus [0, 0.5]^2$ *is to be deformed according to the monitor function*

$$f_{mon}(r) = \min\left\{1, \max\{h \cdot c_0, \sqrt{2}|r|\}\right\}. \tag{3.29}$$

Here, h denotes the mesh size of an equidistant tensor product grid with NEL elements; c_0 acts as shape parameter.

Remark 3.6.3. *The factor of $\sqrt{2}$ is introduced to make the monitor function reach the value 1 in $(-0.5, -0.5)$. Thus the value $h \cdot c_0$ describes the desired fraction of the largest to the smallest element area on the deformed grid. Notice that due to the definition of the monitor function (3.29) the ratio of the largest to the smallest element area grows by a factor of two per regular refinement of the initial grid and thus the difficulties of this test problem grow with the refinement of the starting grid.*

In a first step, we apply our one-level grid deformation algorithm 3.3.1 to this test problem. The shape parameter c_0 is set to 0.1. Like in our previous computations, we solve the deformation IVPs by RK3 with fixed step size of $\Delta t = 0.1$. The deformation PDE is approximated with conforming bilinear Finite Elements and its gradient is reconstructed employing INT. We choose the number of adaptation steps N_a according to formula (3.21) with γ_0 set to 10 like in the previous numerical tests. We perform one correction step after the actual deformation, i.e. we set $N_c(N_a) = 1$.

NEL	N_a	$\alpha_{\min}$	$\alpha_{\max}$	$h_{\min}$	Q_0	Q_∞
768	3	48.15	129.27	$2.97 \cdot 10^{-3}$	$1.07 \cdot 10^{-2}$	$6.17 \cdot 10^{-1}$
3,072	3	25.55	155.42	$7.83 \cdot 10^{-4}$	$5.95 \cdot 10^{-3}$	$1.59 \cdot 10^{0}$
12,288	4	-	-	-	-	-
49,152	4	-	-	-	-	-
196,608	4	-	-	-	-	-

Table 3.6.1: Results for test problem 3.6.2 computed with deformation algorithm 3.3.1 ("$-$" indicates that a test leads to invalid grids)

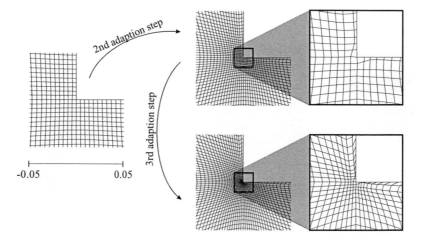

Figure 3.6.1: Excerpts from the grids created by deformation algorithm 3.3.1, $NEL = 12,288$

The results in table 3.6.1 document the failure (indicated by "−" in the table) of our one-level deformation method 3.3.1 for this test problem on fine computational grids. Although the quality measures are fairly low on rather coarse grids, the computation even fails for fine grids as the deformation method creates non-convex elements in the vicinity of the reentrant corner. In figure 3.6.1, we display the intermediate grids for $NEL = 12,288$ after the first, second and third adaptation step. Already after the second adaptation step, the resulting intermediate grid exhibits irregular variations on the element size, as the boundary elements near the reentrant corner are much smaller than their neighbours. This is in contrast to all grids shown previously where the element area distribution is rather smooth. Moreover, the grid points on the domain boundary seem not "to move close enough" to the reentrant corner during the third deformation step which yields large maximal angles in this region of the grid. Because of these local variations, the intermediate grid after the second adaptation step is not a suitable starting point for the third adaptation substep, as the resulting mesh after this step contains non-convex elements in exactly the region the local variations occurred.

Obviously, our grid deformation method does not balance these local mesh perturbations in subsequent deformation substeps. This is confirmed by the fact that increasing N_a by setting a lower value of γ_0 in formula (3.21) leads to non-convex elements as well. Numerical tests reveal that the reported failure *cannot be cured neither by applying more time steps nor by changing the method of reconstructing the deformation vector field*. The reason for the deformation method to fail is evidently the approximate solution of the deformation Poisson problem only, as its accuracy is affected by the reentrant corner. As pointed out above, in the region of the reentrant corner the error is concentrated which corresponds to the region where the deformation method fails. These findings confirm the suspicion that the difficulties we experience are connected with the deformation vector field. Unfortunately, for computing the deformation vector field more accurately we would need either a finer grid which leads to large problems and therefore long computational times or we need a grid which is adjusted to this special situation. To obtain this, however, we perform grid deformation. Thus, we present another remedy.

The insufficient grid quality after a grid adaptation step in deformation process near the reentrant corner can be easily improved by applying certain grid smoothing. Thus, we expect that the grid emerging from the subsequent substep will feature better grid quality, too, as this deformation substep takes place on a smoothed grid. Smoothing of intermediate grids is realised in the improved deformation algorithm 3.4.1 which we investigated in section 3.4. However, smoothing of intermediate grids can only cure the symptoms and does not improve the quality of the Poisson solution itself. Thus, we expect a gain of robustness using the improved deformation algorithm, but still may face failures. We apply grid deformation algorithm 3.4.1, which incorporates postprocessing, to test problem 3.6.2 with the same parameter settings as before. The postprocessing in algorithm

NEL	N_a	$\alpha_{\min}$	$\alpha_{\max}$	$h_{\min}$	Q_0	Q_∞
768	3	52.24	126.06	$3.51 \cdot 10^{-3}$	$1.73 \cdot 10^{-2}$	$6.49 \cdot 10^{-1}$
3,072	3	50.95	127.24	$9.96 \cdot 10^{-4}$	$6.39 \cdot 10^{-3}$	$5.72 \cdot 10^{-1}$
12,288	4	51.22	127.56	$2.89 \cdot 10^{-4}$	$2.18 \cdot 10^{-3}$	$3.05 \cdot 10^{-1}$
49,152	4	46.38	134.00	$9.26 \cdot 10^{-5}$	$8.47 \cdot 10^{-4}$	$3.92 \cdot 10^{-1}$
196,608	4	-	-	-	-	-

Table 3.6.2: Results for test problem 3.6.2 computed with algorithm 3.4.1, postprocessing by 2 Laplacian smoothing steps after every adaptation substep

NEL	N_a	$\alpha_{\min}$	$\alpha_{\max}$	$h_{\min}$	Q_0	Q_∞
768	3	53.88	124.93	$3.51 \cdot 10^{-3}$	$1.81 \cdot 10^{-2}$	$6.56 \cdot 10^{-1}$
3,072	3	53.11	126.06	$9.93 \cdot 10^{-4}$	$6.52 \cdot 10^{-3}$	$5.75 \cdot 10^{-1}$
12,288	4	52.68	126.61	$2.91 \cdot 10^{-4}$	$2.21 \cdot 10^{-3}$	$3.21 \cdot 10^{-1}$
49,152	4	52.44	126.82	$9.60 \cdot 10^{-5}$	$8.18 \cdot 10^{-4}$	$3.13 \cdot 10^{-1}$
196,608	4	-	-	-	-	-

Table 3.6.3: Results for test problem 3.6.2 computed with deformation algorithm 3.4.1, postprocessing with 2 Laplacian smoothing steps and 4 grid optimisation steps

3.4.1 is realised by applying 2 Laplacian smoothing steps. The results presented in table 3.6.2 exhibit the expected gain of robustness and thus confirm the results from Section 3.4. For $NEL = 12,288$, we display the intermediate grids in figure 3.6.2. The comparison with the ones obtained from our first computation without any postprocessing (figure 3.6.1) reveals that the postprocessing by Laplacian smoothing indeed balances the local perturbations in the mesh. Yet on very fine meshes, our deformation method still fails. On these grids, the disturbances by the singularity in the solution of the Poisson problem obviously dominate to such on extent that even when starting from a smoothed grid, one of the adaptation steps produces an invalid grid. Therefore, applying more smoothing steps or a more powerful smoothing algorithm will not further increase the robustness of the deformation. The data in table 3.6.3 are obtained by the same calculation as before, but we realise the postprocessing after each adaptation step by applying 2 Laplacian smoothing steps followed by 4 grid optimisation steps. This combination of Laplacian smoothing and grid optimisation proved well suited as postprocessing in the numerical tests of section 3.4. For the grid optimisation, we choose the same parameters as in this preceding section. As anticipated, employing more powerful postprocessing does not yield enhanced robustness compared to simple Laplacian smoothing.

Overall, obviously even our most advanced one-level deformation algorithm 3.4.1 is not capable to cope with the difficulties arising in test problem 3.6.2.

In the preceding section 3.5, it turned out that the multilevel deformation

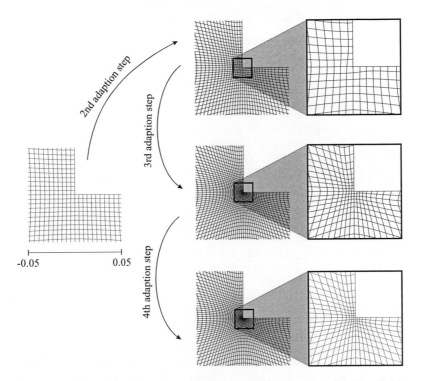

Figure 3.6.2: Excerpts from the grids created by deformation algorithm 3.4.1, postprocessing by 2 Laplacian smoothing steps, $NEL = 12,288$

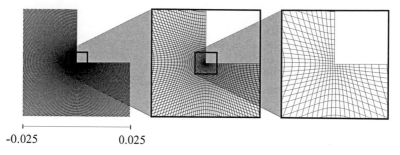

-0.025 0.025

Figure 3.6.3: Excerpts from the grids created by deformation algorithm 3.5.3, $NEL = 196,608$

method is superior to the one-level deformation in many aspects: For fine grids, it is faster, more accurate and more robust. The latter property was deduced from the fact that apparently the multilevel deformation converges for a larger class of problems. As for the L-domain, the difficulties described become the more severe the finer the grid is, it is straightforward to apply the multilevel deformation algorithm to the L-domain. Doing so, we can expect significant speed-up and enhanced accuracy of the deformation which may improve the quality of the emerging grids as well. As a grid produced by exact deformation never contains tangled elements provided that the mesh size is small enough, we can hope that the grids deformed by the multilevel deformation remain admissible as well.

In our multilevel deformation algorithm 3.5.3, we set the starting grid level $i_{min} = 3$ leading to an initial grid consisting of 768 elements. We realize pres-moothing by applying 2 Laplacian smoothing steps and the postprocessing after every adaptation step by applying 3 Laplacian smoothing steps. The level increment i_{incr} is set to one, the number of adaptation steps on every level is again computed by equation (3.21), the blending parameters according to formula (3.26). This leads to 3 adaptation steps on the initial grid level and to one adaptation step on each subsequent one. In contrast to the one-level deformation algorithms considered previously, the multi-level deformation is able to solve the test problem considered on arbitrarily fine grids (table 3.6.4). The number of grid points was restricted by hardware limitations only in this test. The excerpts of the resulting grid for $NEL = 196,608$ in figure 3.6.3 show that the grid generated with the multilevel deformation is free from the local perturbations observed before.

A closer look at table 3.6.4 shows that again we observe that $Q_0 = \mathcal{O}(\sqrt{N})$ and $Q_\infty = \mathcal{O}(\sqrt{N})$. Thus, the multilevel deformation method is *convergent with optimal order* even in the situation of the L-domain. This is particularly remarkable, because almost all preliminaries used in the derivation of the deformation algorithm in section 3.1 are not valid any more: The monitor function depends on the different grids whereas in the analysis, there was one monitor function only

NEL	α_{min}	α_{max}	h_{min}	Q_0	Q_∞
3,072	53.98	125.10	$1.42 \cdot 10^{-3}$	$8.62 \cdot 10^{-3}$	$5.61 \cdot 10^{-1}$
12,288	52.73	126.43	$3.95 \cdot 10^{-4}$	$3.14 \cdot 10^{-3}$	$3.75 \cdot 10^{-1}$
49,152	52.21	127.20	$1.19 \cdot 10^{-4}$	$1.12 \cdot 10^{-3}$	$2.33 \cdot 10^{-1}$
196,608	51.95	127.69	$3.96 \cdot 10^{-5}$	$3.97 \cdot 10^{-4}$	$1.52 \cdot 10^{-1}$
786,432	52.14	127.67	$1.38 \cdot 10^{-5}$	$1.41 \cdot 10^{-4}$	$1.01 \cdot 10^{-1}$
3,145,728	52.07	130.60	$4.75 \cdot 10^{-6}$	$4.99 \cdot 10^{-5}$	$8.36 \cdot 10^{-2}$
12,582,912	43.40	157.82	$1.59 \cdot 10^{-6}$	$1.78 \cdot 10^{-5}$	$1.42 \cdot 10^{-1}$

Table 3.6.4: Results for test problem 3.6.2 computed with the multilevel grid deformation algorithm 3.5.3

for the whole sequence of grids. The monitor functions are not bounded from below by a positive number independent of h as $\min_{x \in \Omega} f(x) = h \cdot c_0$ according to formula (3.29) in contrast to the analysis before. It is continuous, but not in $\mathcal{C}^1(\bar{\Omega})$ as needed for the proof of the convergence theorem 3.1.10. The sequence of deformed grids is *not* size regular in the sense of definition 3.1.2. But all of that, the multilevel deformation algorithm converges with optimal order.

Chapter 4

Application to the Poisson equation

4.1   A priori r-adaptivity on the L-domain

In this section, we shift our attention from our grid deformation method itself to its application to the Poisson problem. We consider test problem 3.6.1 which was introduced in the preceding section and was chosen because its solution features a corner singularity. This leads to the well-known decay in the convergence rate because of the lack of solution regularity. However, this can be balanced by concentrating the mesh around the reentrant corner. For details, we refer to the comprehensive work of Ciarlet [80]. We compare the computations on equidistant meshes with the ones on meshes resulting from test problem 3.6.2. The shape parameter c_0 in the monitor function (3.29) is set to 0.1 as before. We apply multilevel deformation with the same settings as used in section 3.6. Besides the computation of the gradient error by evaluating the difference between the true solution and its FEM counterpart, this error is estimated using both employing the SPR- or the PPR method. The gradient error as well as the corresponding efficiency indices are collected in table 4.1.1 and visualised in figure 4.1.1. In this figure, additionally the gradient errors on meshes deformed according to (3.29) with parameters $c_0 = 0.05$ and $c_0 = 0.01$ are presented. For an illustration of the grids emerging from the deformation method, we refer to section 3.3. It is obvious that the solutions computed on the deformed meshes are far more accurate in terms of the gradient error than the solution computed on the regular mesh. Moreover, using an adapted mesh, it is obviously possible to recover almost the full order of convergence even for non-smooth solutions. In the presented example, we have for $c_0 = 0.01$, $\|\nabla e\| = \mathcal{O}(NEL^\sigma)$ with $\sigma = 0.496$ which is almost optimal in the sense that it is very close to $\sigma = 0.5$, the maximal convergence rate possible with bilinear Finite Elements. This result, however, cannot be interpreted as $\|\nabla e\| = \mathcal{O}(h)$ as for the deformed meshes $NEL \approx h^{-2}$ does not hold any more. As the tests show that the quality of the solution does not depend essentially on the parameter c_0, we consider in the following the case $c_0 = 0.01$ only. Grid details like $h_{\min}$ or the aspect ratio are listed in section 3.6. A closer look at table 4.1.1 shows that by deformation, the quality of error estimation has been improved considerably as well. This phenomenon will be investigated in more detail in the subsequent section.

Remark 4.1.1. *The correct computation of the exact gradient error* $\|\nabla(u-u_h)\|_0$ *turns out to be nontrivial for test problem 3.6.1 due to* ∇u *which exhibits a singularity in* $(0,0)$. *All standard methods for numerical cubature however rely on the smoothness of the function to integrate and thus can not be used here. Because of this, we compute the gradient error not by direct numerical integration. Instead, we proceed following an idea by Babuška. To avoid integrating over the singularity, we exploit that for the Poisson equation, gradient error and energy error coincide:*

$$\|\nabla(u - u_h)\|_0^2 = a(u - u_h, u - u_h) = a(u, u) + a(u_h, u_h) - 2a(u, u_h).$$

Furthermore, we employ that u *is even a strong solution of the Laplace equation on* Ω. *This allows us to perform integration by parts:*

$$a(u, u) = \underbrace{(f, u)}_{=0} + \int_{\partial\Omega} u \cdot \partial_n u \, ds.$$

Another integration by parts yields

$$a(u - u_h, u) = -(\Delta u, u - u_h) + \int_{\partial\Omega} (u - u_h)\partial_n u \, ds$$

and thus we end up with

$$\|\nabla(u - u_h)\|_0^2 = a(u_h, u_h) + \int_{\partial\Omega} u \cdot \partial_n u \, ds - 2\int_{\partial\Omega} \partial_n u \cdot u_h \, ds. \qquad (4.1)$$

The first term in this sum can be computed element by element using standard numerical cubature rules, as on every element T, *it holds* $u_h \in \mathcal{C}^\infty(T)$. *The second term can be treated analytically.*

The last term $\int_{\partial\Omega} \partial_n u \cdot u_h \, ds$ *remains. Although the singularity of* ∇u *is located in* $(0,0) \in \partial\Omega$, *on the corresponding lines* $[(0,0),(0,0.5)]$ *and* $(0,0),(0.5,0)$ *it holds* $u_h = 0$ *due to the Dirichlet boundary conditions. Therefore, the integrand on these lines is zero at all. On the other parts of the boundary, it is again piecewise smooth and can thus be treated using standard 1D cubature methods on the corresponding element edges. The results in table 4.1.2 show that as predicted the cubature error in the direct approach is not of higher order. Because of this, when computing test problem 3.6.1, we evaluate the gradient error using formula (4.1) without stating this explicitly.*

NEL	$\|u - u_h\|_1$	I_{eff} (SPR)	I_{eff} (PPR)	$\|u - u_h\|_1$	I_{eff} (SPR)	I_{eff} (PPR)
3,072	$2.21 \cdot 10^{-2}$	0.726	0.625	$8.11 \cdot 10^{-3}$	0.956	0.921
12,288	$1.40 \cdot 10^{-2}$	0.729	0.629	$4.06 \cdot 10^{-3}$	0.962	0.940
49,152	$8.81 \cdot 10^{-3}$	0.730	0.631	$2.04 \cdot 10^{-3}$	0.966	0.949
196,608	$5.56 \cdot 10^{-3}$	0.731	0.632	$1.03 \cdot 10^{-3}$	0.965	0.952
786,432	$3.51 \cdot 10^{-3}$	0.732	0.633	$5.19 \cdot 10^{-4}$	0.966	0.954
3,145,728	$2.21 \cdot 10^{-3}$	0.732	0.634	$2.63 \cdot 10^{-4}$	0.969	0.959
12,582,912	$1.39 \cdot 10^{-3}$	0.733	0.634	$1.37 \cdot 10^{-4}$	0.972	0.972

Table 4.1.1: Gradient error and efficiency indices for test problem 3.6.1, regular grids (left part) vs. deformed grids (right part), $c_0 = 0.1$

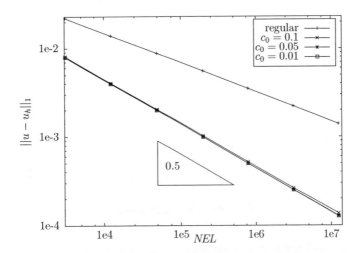

Figure 4.1.1: Gradient errors on regular and deformed grids for different choices of c_0 in formula (3.29)

NEL	$\|u - u_h\|_1$ (Babuška)	$\|u - u_h\|_1$ (direct)
3,072	$2.21 \cdot 10^{-2}$	$2.11 \cdot 10^{-2}$
12,288	$1.40 \cdot 10^{-2}$	$1.34 \cdot 10^{-2}$
49,152	$8.81 \cdot 10^{-3}$	$8.44 \cdot 10^{-3}$
196,608	$5.56 \cdot 10^{-3}$	$5.32 \cdot 10^{-3}$
786,432	$3.51 \cdot 10^{-3}$	$3.36 \cdot 10^{-3}$
3,145,728	$2.21 \cdot 10^{-3}$	$2.12 \cdot 10^{-3}$
12,582,912	$1.39 \cdot 10^{-3}$	$1.33 \cdot 10^{-3}$

Table 4.1.2: Gradient error for test problem 3.6.1, regular grids, evaluation according to equation (4.1) and direct evaluation using the 3×3-Gauss formula (right)

4.2 Analysis of gradient error estimation on the L-domain

In this section, we investigate the two gradient error estimators SPR and PPR with respect to their suitability as basis for r-adaptivity driven by a posteriori control of the gradient error. Our investigations are carried out for the L-domain with the associated test problem 3.6.1. In a fully r-adaptive code, the monitor function has to be computed from the (estimated) error distribution. This requires an error estimator which does not only feature global reliability and efficiency in the sense of equations (1.1) and (1.2) but moreover, the error estimator must give insight into the error distribution. This implies that for $T \in \mathcal{T}$ the local efficiency indices

$$I_{\text{eff},T} := \frac{\eta_T(u_h)}{\|e_h\|_{1,T}}$$

are desired to be close to one like the global efficiency index. As global bounds for the local efficiency indices, we define

$$c := \min_{T \in \mathcal{T}} \{I_{\text{eff},T}\}, \quad C := \max_{T \in \mathcal{T}} \{I_{\text{eff},T}\}.$$

Similar to [82], we introduce the *robustness index*

$$R := \max \left\{ C - 1, \frac{1}{c} - 1 \right\}.$$

Note that the ideal value $R = 0$ implies that the error has been estimated exactly and that moreover the error distribution is known precisely.

Due to the non-smooth nature of the solution of our test problem 3.6.1, we cannot expect that the gradient error is estimated asymptotically exact with the reconstruction-based approaches SPR and PPR. However, the results presented in table 4.1.1 show that during the adaptation process the efficiency index improves significantly for both error estimators and comes very close to one. To explain this phenomenon, we take into account that gradient recovery is a *local* process. Consequently, the local quality of gradient recovery depends on the local smoothness of the underlying solution. For the example of the L-domain, the solution is smooth except of the reentrant corner, where the gradient exhibits a singularity. Hence, on the elements adjacent to the reentrant corner, the gradient recovery partially fails and the local error of these elements cannot be estimated properly. In contrast to this, for all other elements, the solution is smooth and one can expect a good estimation of the local error. These considerations lead to the following theorem.

Theorem 4.2.1. *Let η denote an error estimator for the gradient error with*

$$\eta = \sqrt{\sum_{T \in \mathcal{T}} \eta_T^2}.$$

Assume that $\mathcal{T} = \mathcal{T}_1 \dot\cup \mathcal{T}_2$ with $|\mathcal{T}_2| \leq k$. For $T \in \mathcal{T}_1$ may hold

$$1 - \varepsilon \leq I_{\text{eff},T} \leq 1 + \varepsilon \tag{4.2}$$

with $0 < \varepsilon$ small. For $T \in \mathcal{T}_2$, we assume

$$c \leq I_{\text{eff},T} \leq C. \tag{4.3}$$

Let us denote the ratio of the minimal and maximal error contribution by

$$\omega := \frac{\max_{T \in \mathcal{T}} e_T}{\min_{T \in \mathcal{T}} e_T}. \tag{4.4}$$

Then, we obtain the upper bound

$$I_{\text{eff}} \leq (1 + \varepsilon) \sqrt{1 + \left(\frac{C^2}{(1+\varepsilon)^2} - 1 \right) \frac{k}{NEL - k} \omega^2}.$$

If

$$1 > \left(2 - \frac{c^2}{(1-\varepsilon)^2} \right) \frac{k}{NEL} \omega^2,$$

we obtain the lower bound

$$I_{\text{eff}} \geq (1 - \varepsilon) \sqrt{1 - \left(2 - \frac{c^2}{(1-\varepsilon)^2} \right) \frac{k}{NEL} \omega^2}.$$

Proof. Obviously, for $0 < a, b, \beta$ we have

$$\frac{\sqrt{a + \beta b}}{\sqrt{a + b}} = \sqrt{1 + \frac{(\beta - 1)b}{a + b}} \leq \sqrt{1 + \frac{(\beta - 1)b}{a}}. \tag{4.5}$$

$$I_{\text{eff}} = \sqrt{\frac{\sum_{T \in T} \eta_T^2}{\sum_{T \in T} e_T^2}}$$

$$= \sqrt{\frac{\sum_{T \in T_1} \eta_T^2 + \sum_{T \in T_2} \eta_T^2}{\sum_{T \in T} e_T^2}}$$

$$\overset{(4.2),(4.3)}{\leq} (1+\varepsilon) \sqrt{\frac{\sum_{T \in T_1} e_T^2 + \frac{C^2}{(1+\varepsilon)^2} \sum_{T \in T_2} e_T^2}{\sum_{T \in T} e_T^2}}.$$

$$\overset{(4.5)}{\leq} (1+\varepsilon) \sqrt{1 + \left(\frac{C^2}{(1+\varepsilon)^2} - 1 \right) \frac{k \cdot \max_{T \in T} e_T^2}{(NEL - k) \cdot \min_{T \in T} e_T^2}}.$$

Using (4.4), the proof of the upper bound is finished. With the relation

$$\frac{\sqrt{a + \beta b}}{\sqrt{a + b}} \geq \sqrt{1 + \frac{(\beta - 2)b}{a + b}}, \quad a, b, \beta > 0 \wedge \frac{a}{b} > 1 - \beta,$$

the proof of the lower bound can be obtained in a similar way. □

Corollary 4.2.2. *The assumptions of theorem 4.2.1 may hold. Additionally, let the error estimation on T_1 be asymptotically exact, i.e. $\lim \varepsilon = 0$, $NEL \to \infty$. If $k = |T_2|$, c and C are uniformly bounded in NEL and if $\omega = o(\sqrt{NEL})$, then the error estimation is asymptotically exact.*

Proof. As $\omega^2 = o(NEL)$, the statement follows immediately from theorem 4.2.1. □

Remark 4.2.3. *The assumptions reflect the situation of the L-domain, where T_2 is suspected to be a rather small region around the reentrant corner because of the lack of regularity in this corner. In the case of regular refinement on the L-domain, it is expected that ω is not uniformly bounded in h due to the singularity in the reentrant corner. The boundedness of ω has to be enforced by grid adaptation, e.g. by adaptive refinement or grid deformation.*
However, grid deformation degrades the geometrical regular structure of the grid which is one of the main assumptions in the proofs of superconvergence of the recovered gradient. Therefore, $\varepsilon \to 0$ cannot be stated immediately for deformed grids.

The bounds for the efficiency index presented in theorem 4.2.1 suggest that the quality of error estimation increases if the ratio of the minimal and maximal error contribution decreases, i.e. if the grid is made such that the error contributions are nearer to equidistribution. To investigate the influence of the grid adaptation on the quality of estimating the gradient error, we consider again test problem 3.6.1. The parameter ε in theorem 4.2.1 is set to 0.1, hence in this situation, k denotes the number of elements where the local efficiency index either exceeds 1.1 or falls below 0.9. We deform the grids using the multi-level deformation algorithm

NEL	I_{eff}	ω	C	c	k	R
192	$7.17 \cdot 10^{-1}$	$3.80 \cdot 10^{2}$	8.13	0.48	70	7.13
768	$7.22 \cdot 10^{-1}$	$2.48 \cdot 10^{3}$	11.79	0.49	192	10.79
3,072	$7.26 \cdot 10^{-1}$	$1.54 \cdot 10^{4}$	17.41	0.49	586	16.41
12,288	$7.29 \cdot 10^{-1}$	$9.25 \cdot 10^{4}$	26.32	0.49	1,800	25.32
49,152	$7.30 \cdot 10^{-1}$	$5.37 \cdot 10^{5}$	40.47	0.49	5,361	39.47
196,608	$7.31 \cdot 10^{-1}$	$3.03 \cdot 10^{6}$	62.93	0.49	16,249	61.93
786,432	$7.32 \cdot 10^{-1}$	$1.67 \cdot 10^{7}$	98.58	0.49	50,162	97.58
3,145,728	$7.32 \cdot 10^{-1}$	$9.09 \cdot 10^{7}$	155.18	0.49	156,260	154.18
12,582,912	$7.33 \cdot 10^{-1}$	$5.08 \cdot 10^{8}$	245.19	0.49	491,270	244.19

Table 4.2.1: Results for test problem 3.6.1 using equidistant tensor product grids, $\varepsilon = 0.1$, error estimation with SPR

NEL	I_{eff}	ω	C	c	k	R
192	$7.17 \cdot 10^{-1}$	$3.80 \cdot 10^{2}$	8.13	0.48	70	7.13
768	$9.57 \cdot 10^{-1}$	$3.37 \cdot 10^{2}$	2.35	0.24	131	3.14
3,072	$9.56 \cdot 10^{-1}$	$1.38 \cdot 10^{3}$	2.35	0.21	189	3.82
12,288	$9.62 \cdot 10^{-1}$	$5.25 \cdot 10^{3}$	2.35	0.19	375	4.22
49,152	$9.66 \cdot 10^{-1}$	$2.01 \cdot 10^{4}$	2.33	0.19	783	4.24
196,608	$9.65 \cdot 10^{-1}$	$7.95 \cdot 10^{4}$	2.42	0.19	1,644	4.18
786,432	$9.66 \cdot 10^{-1}$	$3.17 \cdot 10^{5}$	3.71	0.20	3,460	4.08
3,145,728	$9.69 \cdot 10^{-1}$	$1.46 \cdot 10^{6}$	4.64	0.20	7,341	3.94
12,582,912	$9.72 \cdot 10^{-1}$	$8.87 \cdot 10^{6}$	5.56	0.22	14,231	4.56

Table 4.2.2: Results for test problem 3.6.1 computed on grids deformed according to monitor function (3.29), $\varepsilon = 0.1$, $c_0 = 0.1$, error estimation with SPR

3.5.3 with the same settings as in section 3.6. We set the shape parameter in the monitor function to $c_0 = 0.1$.

The results of these numerical tests are presented in table 4.2.1 for undeformed and in table 4.2.2 for deformed meshes, the error estimation was performed using the SPR technique. In order to compare the SPR technique with the PPR method in terms of local error estimation, the numerical tests have been repeated using the PPR method. These results are presented in table 4.2.3 and table 4.2.4. The elementwise integrals occurring in these test were computed by dividing the element into four subelements and applying the 3×3-Gauss cubature rule on each of them.

It turns out that the number of elements on which the local error estimation lacks exactness is significantly reduced by using a deformed grid on higher levels of refinement for both methods of gradient recovery. Moreover, the global quality of the estimation is superior on the highly deformed meshes applied here. This is a rather new aspect as the superior exactness of error estimation using gradient

NEL	I_{eff}	ω	C	c	k	R
192	$6.09 \cdot 10^{-1}$	$3.80 \cdot 10^2$	4.18	0.71	58	3.18
768	$6.19 \cdot 10^{-1}$	$2.48 \cdot 10^3$	5.49	0.71	166	4.49
3,072	$6.25 \cdot 10^{-1}$	$1.54 \cdot 10^4$	7.72	0.71	542	6.72
12,288	$6.29 \cdot 10^{-1}$	$9.25 \cdot 10^4$	11.38	0.69	1,720	10.38
49,152	$6.31 \cdot 10^{-1}$	$5.37 \cdot 10^5$	17.26	0.69	5,272	16.26
196,608	$6.32 \cdot 10^{-1}$	$3.03 \cdot 10^6$	26.66	0.68	16,111	25.66
786,432	$6.33 \cdot 10^{-1}$	$1.67 \cdot 10^7$	41.59	0.68	49,935	40.59
3,145,728	$6.34 \cdot 10^{-1}$	$9.09 \cdot 10^7$	65.59	0.67	155,917	64.59
12,582,912	$6.34 \cdot 10^{-1}$	$5.08 \cdot 10^8$	108.87	0.68	490,817	107.87

Table 4.2.3: Results for test problem 3.6.1 for equidistant tensor product grids, $\varepsilon = 0.1$, error estimation with PPR

NEL	I_{eff}	ω	C	c	k	R
192	$6.09 \cdot 10^{-1}$	$3.80 \cdot 10^2$	4.18	0.71	58	3.18
768	$9.03 \cdot 10^{-1}$	$3.37 \cdot 10^2$	2.86	0.55	83	1.86
3,072	$9.21 \cdot 10^{-1}$	$1.38 \cdot 10^3$	2.77	0.53	75	1.77
12,288	$9.40 \cdot 10^{-1}$	$5.25 \cdot 10^3$	2.71	0.50	143	1.71
49,152	$9.49 \cdot 10^{-1}$	$2.01 \cdot 10^4$	3.03	0.50	275	2.03
196,608	$9.52 \cdot 10^{-1}$	$7.95 \cdot 10^4$	3.69	0.51	535	2.69
786,432	$9.54 \cdot 10^{-1}$	$3.17 \cdot 10^5$	3.63	0.51	1,072	2.63
3,145,728	$9.59 \cdot 10^{-1}$	$1.46 \cdot 10^6$	2.32	0.48	2,161	1.32
12,582,912	$9.72 \cdot 10^{-1}$	$8.87 \cdot 10^6$	3.13	0.40	4,460	2.13

Table 4.2.4: Results for test problem 3.6.1 using grids deformed according to monitor function (3.29), $c_0 = 0.1$, $\varepsilon = 0.1$, error estimation with PPR

recovery is strongly related to the superconvergence of the recovered gradient itself. For the overwhelming majority of the superconvergence proofs found in literature, geometric regularity of the grid besides of high regularity of the solution appear as essential ingredients. In the situation of the L-domain, the solution is neither smooth nor the grid geometrically regular.

However, on coarse grids the number of "bad" elements for deformed meshes exceeds the number of "bad" elements on the regular mesh indicating that the geometric regularity of the grid does indeed play a role. Moreover, although the quality of estimation is very good on deformed meshes, there is no indication for asymptotic exactness of estimation and therefore no evidence of superconvergence. The numerical tests exhibit that k is apparently *not* bounded in any of the cases investigated here in spite of

$$\left|\left\{T \big| u|_{\bar{T}} \notin H^{\infty}\right\}\right| = 12$$

holds on all grids applied.

Furthermore, the numerical tests exhibit that by deformation c and C become bounded which is not the case on regular grids. This indicates in junction with the robustness index R that appropriate grid deformation can considerably improve the estimation of local error contributions. Both for the deformed and the undeformed grid, ω is not bounded. However, on the undeformed mesh, ω increases with an average factor of 6.03 per level of refinement which corresponds to $\omega = \mathcal{O}(NEL^{3/2})$. This factor reduces to 4.42 on the deformed mesh (both values for PPR-method). This leads to the conclusion that there is room for improvement in choosing an appropriate monitor function.

In summary, the investigation of the two methods of gradient recovery reveals that both methods lead to reasonable estimations of the gradient error and its distribution in the domain. Therefore, both methods are suitable building blocks for the r-AFEM to develop in the next section. However, as SPR requires less computational effort than PPR, we concentrate on this method in what follows.

4.3 Gradient error driven r-adaptivity on the L-domain

In the following, we aim at incorporating the deformation method in a fully adaptive Finite Element method for the Poisson equation. In contrast to the previous section, the monitor function f is now generated from the estimated error distribution $\eta(x)$ computed in the preceding step in the adaptive iterative loop. This section is intended to investigate in brief the overall feasibility of our deformation method as part of an r-adaptive FEM (r-AFEM). It does not pretend to provide an exhaustive analysis, either theoretical or numerical, of r-adaptive FEM, as this analysis, in particular the analysis of convergence aspects, is a whole field of research at its own.

Analogously to the well-known h-AFEM ([37]), we define the r-AFEM we will employ by the following generic algorithm.

Algorithm 4.3.1 (generic r-AFEM).

input: • $GRID$: initial computational grid
          • J: target functional
          • TOL: error tolerance
          • f, g: right hand side and boundary data
          • $i_{\max}$: maximal number of adaptive iterations

output: • $J(u_h)$: approximate value of target functional
          • η: estimated error

function r-**AFEM**$(GRID, J, TOL, f, g, i_{\max}) : J(u_h), \eta$
    $GRID_1 := GRID$
    DO $i = 1, i_{\max}$
        $u_i := $ **SOLVE**$(f, g, GRID_i)$
        $\eta_i := $**ESTIMATE**$(u_i, J)$
        IF $(\eta_i < TOL)$ EXIT LOOP
        $f_{mon,i} := $**MON**$(\eta_i)$
        $GRID_{i+1} := $ **DEFORM**$(f_{mon,i}, GRID_i)$
        IF $(\exists$ non-convex elements$)$ RETURN

    END DO
    $J(u_h) := J(u_i); \eta := \eta_i$
    RETURN $J(u_h), \eta$
END r-**AFEM**

For the gradient error estimation on the L-domain, the four generic building blocks in algorithm 4.3.1 are realised as follows. The first one, **SOLVE**, refers to the solve of the Poisson problem 3.6.1 on the L-domain using the ScaRC-solver of the FEAST package. We will not investigate **SOLVE** further here but refer instead to C. Becker's [12] and S. Kilian's [54] PhD theses for details regarding ScaRC. The block **ESTIMATE** is defined as gradient error estimation using the SPR- or PPR technique. The corresponding analysis was provided in the preceding section 4.1. The last ingredient in algorithm 4.3.1, **DEFORM**, is realised by the multilevel deformation algorithm 3.5.3 which has been investigated in detail in this thesis. In the generic algorithm, the additional parameters of the deformation algorithm 3.5.3 are omitted for convenience. The sanity check for non-convex elements is motivated by the fact that such element can occur due to numerical inaccuracies even though this is theoretically impossible for an exact calculation on sufficiently fine grids.

It remains to define the building block **MON**, which symbolises the construction of a suitable monitor function for the deformation algorithm from the estimated error distribution. From now on, we assume that the error distribution $\eta(x)$ is defined as bilinear FEM function. We identify in the following the elementwise error distribution with its bilinear interpolant. The construction of the monitor function can be formalised by applying an operator M to $\eta(x)$. The operator M plays the same role in r-adaptivity as the marking strategy in h-adaptivity. Finding an optimal marking strategy is not trivial (compare e.g. [37]). This gives a hint that constructing M is non-trivial either. The monitor operator M has to fulfil a number of side conditions. As its output is to serve as monitor function, M must fulfil the condition $M(\eta)(x) > 0 \,\forall x \in \Omega$. Moreover, the smoothness requirements of the grid deformation method lead to $M(\eta) \in \mathcal{C}^1(\Omega)$. However, our numerical experiments in the previous chapters show that for monitor functions in $\mathcal{C}(\Omega)$, all variants of the deformation method work fine. Therefore, we relax the aforementioned condition on $M(\eta) \in \mathcal{C}(\Omega)$. If on a given grid the error is equidistributed, no further deformation should occur, as the grid is already optimal. This leads to $M(c)(x) = \text{area}(x)$, where c represents a constant error distribution. In regions where the error is large, the grid must be concentrated to decrease the error by smaller element sizes. Vice versa, regions with extremely small error can be coarsened without affecting the overall error. Let x_{max} be a point where the error distribution is maximal, and $x_{\min}$ a minimal point, respectively. Then, we require $M(\eta(x_{max})) = \min_{x \in \Omega} M(\eta(x))$ and $M(\eta(x_{\min})) = \max_{x \in \Omega} M(\eta(x))$. Although these conditions present a certain framework, there is still freedom of choice.

As mentioned before, the analogous building blocks to the construction of the monitor function and grid deformation in h-adaptivity are the marking strategy and the technique of local grid refinement or coarsening. This leads to additional conditions to M. In h-adaptive codes, usually the elements marked for refinement are refined or coarsened only once despite the error indicator might suggest performing several refinement or coarsening steps due to the local error size [7]. This strategy may increase the number of adaptive iterations but heuristically it makes the adaptation process more stable. Refining the grid once means decreasing the element area by a factor of four whereas the coarsening implies enlarging the element size by a factor of 4. Thus, starting from a grid with equidistributed element size, for the h-adapted grid after one adaptation step, it holds $\text{area}_{\max} \leq 16\text{area}_{\min}$. If the same is to achieve by grid deformation, the monitor function f has to fulfil $\max_{x \in \Omega} f(x) \leq 16 \min_{x \in \Omega} f(x)$. Because of these considerations, we require M to fulfil the condition $\max_{x \in \Omega} M(\eta)(x) \leq c_M \max_{x \in \Omega} M(\eta)(x)$ with c_M being in the range of ten. In h-adaptivity employing hanging nodes, usually isolated elements which are refined or coarsened are avoided. This implies that the area distribution should not feature local isolated peaks or sinks. Therefore, it is reasonable to consider smoothing of the monitor function.

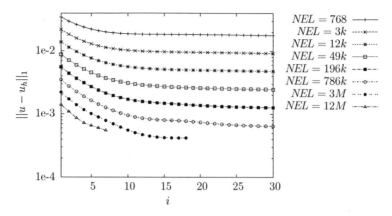

Figure 4.3.1: Gradient error vs. number of r-adaption steps i for different levels of refinement, r-AFEM 4.3.1 applied to test problem 3.6.1

All these considerations lead us to the following definition of M. It should be emphasized that possibly there are many reasonable choices of M.

Definition 4.3.2. *Let* $v \in C(\Omega)$ *be a positive FEM function with* $\max_{x \in \Omega} |v(x)| < 1$. *Then,* c_1 *and* c_2 *being positive constants, we define*

$$M(v)(x) := [-c_1 \ln(\mathbf{Smooth}_i(v(x))) + c_2] \cdot \mathrm{area}(x).$$

Here, $\mathbf{Smooth}_i$ *stands for applying* i *Laplacian smoothing steps to* v.

We now consider the L-domain with test problem 3.6.1 computed with algorithm 4.3.1, $i_{\max} = 30$. We set $c_1 = 1$, $c_2 = 0$ and $i = 0$ in definition 4.3.2. Thus, no smoothing is applied to the monitor function. We estimate the gradient error by SPR. The grids are deformed using the multilevel algorithm 3.5.3 with minimal deformation level $i_{\min} = 2$. This corresponds to a coarse grid consisting of 768 elements. After every level increment in the multilevel deformation, we smooth the grid by two Laplacian smoothing steps. We again employ 10 RK3 steps for ODE solving, the deformation vector field is obtained by using the reconstruction technique INT. The tolerance TOL for the quality measures is set such that no correction step occurs. In figure 4.3.1, we present the history of the true gradient error for these settings and various levels of grid refinement. On the grids with up to $786,000$ elements, the method achieves a steady state after a certain number of iterations. This is indicated by the gradient error approaching a limit. On the fine grids with 3 and 12 millions of elements, the r-adaptive algorithm 4.3.1 leads to non-convex elements after a few adaptive iterations and thus fails to reach the steady state observed on the coarser grids. As the number of possible adaptive iterations decreases from 18 to 7, we expect that on an even finer grid, after less than 7 adaptive iterations non-convex elements will occur. Due to hardware constraints, this conjecture can not be validated. Figure 4.3.2 depicts the exact gradient error on the last computed valid grid for the different levels of refinement. For the sake of comparability, we display additionally the exact gradient error for a sequence of regular tensor product meshes ("regular") and the sequence of grids which stem from the solution of test problem 3.6.2 with the multilevel deformation method ("a priori") These grids can be regarded as deformed a priori. On the grids up to $786,000$ elements produced by our r-AFEM, the gradient error decreases with optimal order although its absolute size is slightly larger than the error on the a priori deformed grids (compare section 4.1). This indicates that the grids created by the r-AFEM 4.3.1 are not optimal. On the grid with 3 millions of elements, the reduction of the gradient error however is lowered, the gradient error on the grid with 12 millions of elements is even larger than the one on the grid with 3 millions of elements. Clearly, this behaviour is related to the inability of the r-AFEM to reach a steady state on these grids. For the sequence of grids with 3 millions of elements, we present the minimal and maximal angle in figure 4.3.3. The minimal angle tends to 0 and the maximal angle to $\pi/2$. This tendency explains the occurrence of the non-convex elements observed and therefore prevents our r-AFEM to reach a steady state. Obviously this is related to the main weakness of our grid deformation method which became apparent in the tests of our one-level deformation on the L-domain (section 3.6): the inability to control the element quality in terms of e.g. aspect ratio or minimal/maximal angle. The observations described motivate the need for postprocessing the grid in order to improve the grid quality after a deformation step.

In another series of tests, we test the benefit of grid postprocessing after the application of the building block **DEFORM** in the r-AFEM 4.3.1. We repeat the computations on the L-domain from above, but apply 2 Laplacian smoothing steps and 4 grid optimisation steps (compare section 3.4) after every level increment and after every one-level deformation step inside the multilevel deformation algorithm. Moreover, we set $i = 2$ in the definition 4.3.2 of M, i.e. the error distribution is treated by 2 Laplacian smoothing steps in the construction process of the monitor function. We show the developing of the gradient errors vs. the number of r-adaptive iterations i in figure 4.3.4 for different levels of refinement. The comparison with the analogous figure 4.3.1 for the former computation reveals the benefit of postprocessing. For the grid with 3 millions of elements, the r-AFEM does not produce invalid grids at all unlike before, for the grid with 12 millions of elements, 17 passes of the iterative loop are performed before at least one element becomes non-convex compared with only 7 without postprocessing. In figure 4.3.5, we plot the gradient error on the last computed admissible grid with ("r-AFEM + pp") and without postprocessing ("r-AFEM"). For convenience, we additionally display the gradient error on regular and a priori deformed grids (see section 3.6). The gradient errors on the grids created by r-AFEM combined with

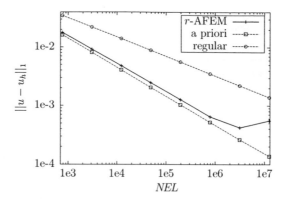

Figure 4.3.2: Gradient error on the final admissible grids vs. number of elements NEL, r-AFEM 4.3.1 applied to test problem 3.6.1

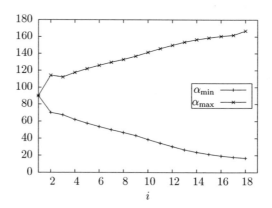

Figure 4.3.3: Minimal and maximal angle α_{min} and α_{max} on the sequence of grids constructed by the r-AFEM 4.3.1 vs. number of r-adaption steps i. The initial grid consists of 3,145,728 elements. The r-AFEM 4.3.1 is applied to test problem 3.6.1.

NEL	I_{eff}, r-AFEM	I_{eff}, reg.
3,072	0.9802	0.7591
12,288	0.9866	0.7614
49,152	0.9916	0.7628
196,608	0.9952	0.7638
768,432	0.9977	0.7643
3,145,728	1.000	0.7647
12,582,912	1.025	0.7649

Table 4.3.1: Efficiency indices on the last computed admissible grids computed with r-AFEM 4.3.1 combined with postprocessing (left) and on unadapted regular grids (right), test problem 3.6.1

postprocessing are smaller than the ones on the grids created without postprocessing for all levels of refinement. In particular, the gradient error decreases on the two finest grids with almost optimal order in contrast to the computation without postprocessing. On the sequence of grids emerging from the r-AFEM 4.3.1 with postprocessing, the gradient error is of order $\mathcal{O}(NEL^{0.489})$ which is very close to the optimal convergence rate of $\mathcal{O}(NEL^{1/2})$. For the computation of this convergence rate, the gradient error on the grid with 12 millions of unknowns was ignored.

Our r-AFEM relies on the error distribution which we estimate by the SPR-method in the tests discussed here. Thus, it is crucial that the estimated error is close to the real one. In table 4.3.1, we list the efficiency indices I_{eff} for the last admissible grid created by the r-AFEM combined with the postprocessing described above. For comparison, the efficiency index for the regular grids is shown in the same table. The efficiency index on the r-adapted grids is very close to one although the geometrical regularity of the grid, which is one of the main ingredients for the superconvergence analysis of SPR, is destroyed by grid deformation. These results are similar to the ones gained by deforming the grids by a given monitor function, see section 4.1. Exchanging SPR by PPR in the r-AFEM 4.3.1 leads to almost the same results regarding both gradient error and efficiency index as the ones shown. Therefore, we omit the presentation of these data.

However, even if postprocessing stabilises and improves the r-AFEM 4.3.1, the method leads to non-convex grids after some adaptive steps on the highest level of refinement considered here. This behaviour reminds of the results for test problem 3.6.2, the deformation on the L-domain with respect to a given monitor function. In this case, postprocessing improves the results but does not provide the ultimate solution of this test problem (compare section 3.6). The multilevel deformation method 3.5.3 proved to be superior for this test setting and only its application led to satisfactory results. These consideration motivate the extension of our pure r-AFEM 4.3.1 to a satisfactory multilevel r-AFEM analogously to multilevel

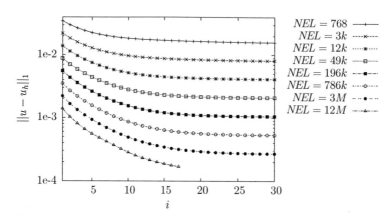

Figure 4.3.4: Gradient error vs. number of r-adaption steps i for different levels of refinement, r-AFEM 4.3.1 with postprocessing applied to test problem 3.6.1

deformation which is in fact an rh-AFEM. This kind of adaptive algorithm we investigate in section 5.2.

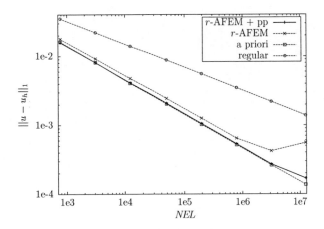

Figure 4.3.5: Gradient error on the last computed admissible grids vs. number of elements *NEL*, *r*-AFEM 4.3.1 with postprocessing applied to test problem 3.6.1

Chapter 5

Outlook: h - and rh-AFEM

5.1 Numerical analysis of h-AFEM on the L-domain

We expounded in the introduction that there are several techniques of grid adaptation which preserve the local logical structure of the grid and thereby are particularly suited in the FEAST context. Besides the well-investigated r-adaptivity, we mulled macrowise h-adaptivity and anisotropic refinement as a sort of a priori h-adaptivity. In this section, we investigate these h-adaptive techniques and compare them in a second step with the r-AFEM developed in the preceding section 4.3 on the L-domain and the associated test problem 3.6.1. Moreover, we provide the gradient error for the same test problem computed with an elementwise h-adaptive algorithm for comparison. These results are calculated by Rademacher [64] with the FEM package SOFAR [15].

For the selection of the macros to refine in macro-wise h-adaptivity, we apply the fixed fraction strategy (compare subsection 1.2.1). The macros are sorted due to the size of their error contributions η_M,

$$\eta_M^2 := \sum_{T \in M} \eta^2(T),$$

and according to equation (1.9), all macros M_i from the beginning up to an index N_1 are marked for refinement. This index is chosen such that

$$\sum_{i=1}^{N_1} \eta_{M_i}^2 \leq \chi \eta^2, \quad 0 < \chi < 1,$$

holds. For macro-wise h-adaptivity, we compute this test problem on two equidistant tensor product meshes consisting of 48 and 768 macros with $\chi = 0.75$. The results (see table 5.1.1) clearly show that macro-wise h-adaptivity does not lead to optimal results for the L-domain. After the initial phase of the computation, where the error is reduced rapidly, we observe the same decay of the convergence

NEL	$\|u - u_h\|_1$	I_{eff}	$l_{\min}$	$l_{\max}$
3,072	$2.21 \cdot 10^{-2}$	0.748	1	1
3,108	$1.55 \cdot 10^{-2}$	0.784	1	2
3,336	$1.06 \cdot 10^{-2}$	0.824	1	3
4,500	$7.35 \cdot 10^{-3}$	0.854	1	4
9,804	$5.04 \cdot 10^{-3}$	0.878	1	5
31,884	$3.43 \cdot 10^{-3}$	0.896	1	6
194,940	$2.32 \cdot 10^{-3}$	0.910	1	7
384,216	$1.56 \cdot 10^{-3}$	0.922	1	8
1,463,436	$1.04 \cdot 10^{-3}$	0.931	1	9
5,720,652	$6.91 \cdot 10^{-4}$	0.938	1	10

Table 5.1.1: Exact gradient error, efficiency indices and minimal and maximal level $l_{\min}$ and $l_{\max}$ for test problem 3.6.1, macro-wise h-adaptive refinement, 768 macros, $\chi = 0.75$

order as for regular refinement (compare figure 5.1.2). It is well-known [80] that the computational grid must be concentrated around the reentrant corner in order to balance the singularity in the solution which causes the order decay. However, applying pure h-adaptivity, the macros remain classical tensor product grids with equidistant mesh size. This implies that around the reentrant corner the grids created by macrowise h-adaptivity are regular grids in fact which explains the decay in convergence order observed. However, due to the rather coarse elements far away from the singularity where the error is small, these grids allow more accurate computations than regular ones with a comparable number of elements. Moreover, by the condition that the difference in the level of refinement must not exceed one for edge-adjacent macros, the overall difference in levels of refinement is bounded to 4 for the grid with 48 macros. If this maximal level is reached during the computation and the macro which features this maximal level is to be refined again, necessarily *all* other macros are refined in the worst case, too. Therefore, in fact a regular refinement step is performed then. Due to the finer granularity in h-adaptivity, the gradient error for the grids consisting of 768 macros is smaller than for the ones with 48 macros only, but for both cases, the same decay of the convergence order can be observed.

For both h-adapted grids, the gradient error is slightly underestimated by SPR, although the efficiency indices are closer to one than for regular refinement (compare table 4.1.1 for regular grids with 768 macros). Exchanging SPR by PPR does not lead to significantly different results, therefore we omit the presentation of these data. Overall, macrowise h-adaptivity can not be regarded as suitable method of grid adaptation for the test example considered and thus will not be investigated more in detail in this thesis.

To perform anisotropic refinement, we modify the subdivision rule in the grid generation algorithm. By default, a quadrilateral element is refined in four by introducing the midpoints of its edges as new vertices. Then, the new vertex in the element center is defined as the intersection of the lines connecting these edge midpoints. The coordinates of the element midpoints can be easily obtained by computing the arithmetic mean of the coordinates of the corresponding vertices. For anisotropic refinement on a single macro, we reduce our considerations for the ease of presentation to the interval $\overline{v_1 v_2}$. For anisotropic refinement in direction of v_1, we parametrise this interval by $[0, 1]$ and define the parameter value t_{new} of the new grid point by the *weighted mean* of the parameters 0 and 1, i.e.

$$t_{\text{new}} := \alpha$$

using the *anisotropy factor* $0 < \alpha < 1$. In the special case $\alpha = 0.5$, isotropic refinement is recovered. After one isotropic refinement step, the midpoint of the leftmost subinterval is then moved in the same way as the former midpoint. This process is repeated until the desired level of refinement is reached. As a consequence, we end up with *graded meshes* (for details, see [80]) which permit concentrations of the elements along the boundary or, in the case of the L-domain, in the vicinity of the reentrant corner. In figure 5.1.1, a grid constructed by anisotropic refinement is displayed.

For the computation of test problem 3.6.1, we start with an equidistant tensor product consisting of 3 macros. The macro edges being part of a coordinate axis are marked for anisotropic refinement with $\alpha = 0.25$. The resulting gradient errors as well as the maximal aspect ratio and the minimal mesh width are displayed in table 5.1.2. It turns out that for anisotropic refinement, the solution converges with almost optimal order. Therefore, the application of anisotropic refinement leads to grids well-adapted to the test problem on the L-domain. However, anisotropic refinement has two severe disadvantages: It lacks flexibility, as it is restricted to boundary contours or to the concentration of elements along a certain line. Moreover, anisotropic refinement implies large aspect ratios in the grid. This can be desirable, e.g. for the approximation of boundary layers, but in the case of the L-domain, it leads to extremely narrow elements in the grid, which are no way adjusted to the underlying problem. Solving on such grids turns out to be challenging for iterative solvers. Additionally, the anisotropy parameter has to be set by the user which requires a priori knowledge about the nature of the solution. Even if in the rather simple case of the L-domain, there are theoretical results giving hints about the optimal degree of mesh grading [80], in more complex situations the user has to rely on his intuition and experience.

Figure 5.1.2 allows a direct comparison of the macrowise h-adaptivity ("mac h") investigated with elementwise h-adaptivity ("el h") and our r-AFEM 4.3.1 ("r-AFEM + pp") as well as anisotropic refinement ("aniso"). It turns out that (at least for the L-domain) r-adaptivity and anisotropic refinement perform by far superior compared to macrowise h-adaptivity. Macrowise h-adaptivity obviously does not offer the flexibility needed for this test example. Although anisotropic refinement provides very accurate results and in particular features an almost optimal convergence order, it requires a priori knowledge about the solution and is

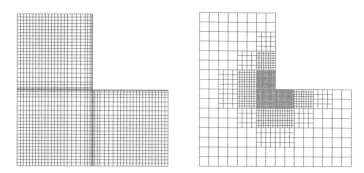

Figure 5.1.1: Anisotropic (left) and macrowise h-adaptive (right) grid on the L-domain

NEL	$\|u - u_h\|_1$	I_{eff} (SPR)	$h_{\min}$	asp. ratio
768	$1.59 \cdot 10^{-2}$	1.086	$1.95 \cdot 10^{-3}$	24
3,072	$8.11 \cdot 10^{-2}$	1.055	$4.88 \cdot 10^{-4}$	48
12,288	$4.10 \cdot 10^{-3}$	1.034	$1.22 \cdot 10^{-4}$	96
49,152	$2.07 \cdot 10^{-3}$	1.022	$3.05 \cdot 10^{-5}$	192
196,608	$1.04 \cdot 10^{-3}$	1.014	$7.63 \cdot 10^{-6}$	384
786,432	$5.21 \cdot 10^{-4}$	1.009	$1.91 \cdot 10^{-6}$	768
3,145,728	$2.61 \cdot 10^{-4}$	1.005	$4.77 \cdot 10^{-7}$	1,536
12,582,912	$1.31 \cdot 10^{-4}$	1.003	$1.19 \cdot 10^{-7}$	3,072

Table 5.1.2: Gradient error, efficiency indices, minimal mesh width and aspect ratio for test problem 3.6.1, anisotropic refinement with $\alpha = 0.25$

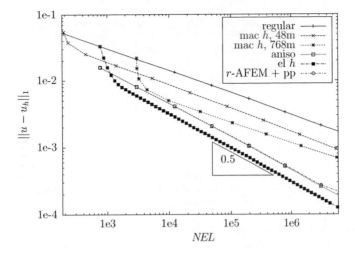

Figure 5.1.2: Comparison of h-adaptive, r-adaptive and anisotropic refinement for test problem 3.6.1

likely to fail in more complex scenarios. Therefore, we do not consider neither macrowise h-adaptivity nor anisotropic refinement any further. The meshes obtained by elementwise h-adaptivity provide for a given number of unknowns the most accurate solution of all types of grids considered here. However, these grids are unstructured and thus are not suitable for high performance computing with FEAST. In contrast to this, our r-AFEM leads relying on a posteriori information only to local logical tensor product grids which provide optimal convergence rates for the L-domain.

5.2 Combining h and r-adaptivity

The test examples considered so far indicate that grid deformation is a viable tool for a posteriori grid adaptation in FEM simulations. However, r-adaptivity aims at relocating the grid points without adding new ones. Therefore, the initial grid has the same number of unknowns as the final grid and thus even the first iteration steps are performed on a fine mesh. This results in high computational cost for every adaptive iteration step. In contrast to this, h-adaptive algorithms start from a coarse mesh and insert as many new grid points as needed during the program run. Because of this, the initial grid for h-adaptive computations is usually very coarse and therefore the first few adaptive iterations steps can be performed very fast. These considerations indicate that regardless of the effort for the grid deformation itself, h-adaptive methods will consume far less computational time than r-adaptive algorithms given the same number of unknowns on the final mesh.

In the context of a-posteriori error control and adaptivity, applying r-adaptivity can be regarded as solving a special optimisation task: The location of the points has to be changed such that the error in the desired output quantity is minimised while preserving the mesh topology. Even if the r-AFEM is capable to provide the optimal mesh, it may happen that the error obtained on this optimal grid still exceeds the prescribed error tolerance. In this case, further passes in the r-adaptive loop are pointless. In practical computations, one has to stop the computation at this point and has to create a new, finer mesh on which the calculations are continued.

Thus, we aim to utilise our r- and h-adaptive method as building blocks for rh-adaptivity such that the advantages of both methods are combined. In the following we describe and investigate two ways of performing rh-adaptivity.

The first rh-algorithm uses r-adaptivity as *"geometric preconditioner"* for the adaptive iteration. To do so, we employ pure r-adaptivity on the coarse grid. If in the n-th r-adaptive step the error η_n does not decrease significantly any more compared to the estimated error η_{n-1} of the previous adaptive iteration step, i.e. $\eta_n > c_r \cdot \eta_{n-1}$, it is reasonable to assume that further r-adaptive iterations will not decrease the error significantly any more. In this case, we proceed with macro-wise h-adaptivity. Doing so, there is no grid deformation on fine grids and thus the computational effort for deforming the grid can be neglected with respect to the overall computational cost. Moreover, we compute our actual problem only once on the finest grid like for pure h-adaptivity. We formulate this kind of rh-AFEM in algorithmic form.

Algorithm 5.2.1 (rh-AFEM 1).

input:
- *GRID: initial computational grid*
- *J: target functional*
- *TOL: error tolerance*
- *f, g: right hand side and boundary data*
- *$i_{\max,r}$: maximal number of r-adaptive iterations*
- *$i_{\max,h}$: maximal number of h-adaptive iterations*
- *c_r: reduction factor for r-AFEM*

output:
- *$J(u_h)$: approximate value of target functional*
- *η: estimated error*

function rh-**AFEM 1**$(GRID, J, TOL, f, g, i_{\max,r}, i_{\max,h}, c_r)$: $J(u_h), \eta$

    $\eta_0 := \infty$

    $GRID_1 := GRID$

    $u_1 := $ **SOLVE**$(f, g, GRID_1)$

    $\eta_1 := $ **ESTIMATE**(u_1, J)

    DO $i = 1, i_{\max,r}$

        IF $(\eta_i < TOL)$ THEN

            $J(u_h) := J(u_i); \eta := \eta_i$

            RETURN $J(u_h), \eta$

        END IF

        IF $(\eta_i > c_r \eta_{i-1})$ EXIT LOOP

        $f_{mon,i} := $**MON**$(\eta_i)$

        $GRID_{i+1} := $**DEFORM**$(f_{mon,i}, GRID_i)$

        IF $(\exists$ *non-convex elements*$)$ RETURN

        $u_{i+1} := $ **SOLVE**$(f, g, GRID_{i+1})$

        $\eta_{i+1} := $ **ESTIMATE**(u_{i+1}, J)

    END DO

    DO $j = i + 1, i + i_{\max,h} + 1$

        MARK*$(\eta_j, GRID_j)$*

        $GRID_{j+1} := $**REFINE**$(GRID)$

        $u_j := $**SOLVE**$(f, g, GRID_j)$

        $\eta_j := $**ESTIMATE**$(u_j, J)$

        IF $(\eta_j < TOL)$ EXIT LOOP

    END DO

    $J(u_h) := J(u_j); \eta := \eta_j$

    RETURN $J(u_h), \eta$

END rh-**AFEM 1**

The second rh-AFEM is motivated by the following considerations. It is well known that for a given number of unknowns, the computational grid is optimal if the error is equidistributed, i.e. if all elementwise contributions are of the same magnitude (for investigations on the L-domain, compare section 4.2). If the error during r-adaptivity does not decrease significantly any more, we can conclude that the grid points of the deformed grid are almost optimally distributed. This leads to the heuristical assumption that the error on this grid is almost equidistributed.

Therefore, it does not make sense to refine parts of the grid using hanging nodes in such situation, but it is necessary to refine the whole grid on time in a regular manner. On this refined grid, the error does not have to be equidistributed any more, such that r-adaptivity on this refined grid may improve the quality of the solution. Note that in this algorithm hanging nodes are avoided entirely.

Algorithm 5.2.2 (rh-AFEM 2).

input:
- *GRID: initial computational grid*
- *J: target functional*
- *TOL: error tolerance*
- *f, g: right hand side and boundary data*
- $i_{\max,r}$: *maximal number of r-adaptive iterations*
- $i_{lev,\min}$: *minimal level for r-AFEM*
- $i_{lev,\max}$: *maximal level for r-AFEM*
- c_r: *reduction factor for r-AFEM*

output:
- $J(u_h)$: *approximate value of target functional*
- η: *estimated error*

function rh-**AFEM 2**$(GRID, J, TOL, f, g, i_{\max,r}, i_{lev,\min}, i_{lev,\min}, c_r)$:
$J(u_h), \eta$

 $\eta_0 := \infty$

 $GRID_1 := GRID$

 DO $i = i_{lev,\min}, i_{lev,\max}$

  DO $j = 1, i_{\max,r}$

   $u_j := $ **SOLVE**$(f, g, GRID_j)$

   $\eta_j := $ **ESTIMATE**(u_j, J)

   IF $(\eta_j < TOL)$ **THEN**

    $J(u_h) := J(u_j); \eta := \eta_j$

    RETURN $J(u_h), \eta$

   END IF

   IF $(\eta_j > c_r \eta_{j-1})$ **EXIT LOOP**

   $f_{mon,j} := $**MON**$(\eta_j)$

   $GRID_{j+1} := $**DEFORM**$(f_{mon,j}, GRID_j)$

   IF $(\exists \text{ non-convex elements})$ **RETURN**

  END DO

  $GRID_1 := $ **PROLONGATE**$(GRID_j)$

 END DO

 $J(u_h) := J(u_j); \eta := \eta_j$

END rh-**AFEM 2**

To compare these two algorithms with pure r-adaptivity, we consider again test problem 3.6.1. The gradient error is estimated using SPR gradient recovery. For both rh-AFEMs to test, the parameter settings described below coincide. The monitor function is created applying the operator M from definition 4.3.2 with $c_1 = 1, c_2 = 0$ and $i = 2$. The grids are deformed using the multilevel grid deformation algorithm 3.5.3, where we set $i_{incr} = 1$ and $i_{min} = 2$. Thus, the coarsest deformed grid consists of 768 elements. The IVPs in the one-level deformation inside the multilevel deformation are solved using 10 equidistant RK3 steps, the deformation vector fields are computed using INT. After every one-level deformation step and after every level increment inside multilevel deformation, we perform 2 Laplacian smoothing steps followed by four grid optimisation steps. These are the same deformation settings which we employed during the tests of the r-AFEM 4.3.1. For the rh-AFEM 5.2.2, we additionally smooth the grid by two Laplacian smoothing steps after every increment of the multigrid level of the actual computation. The maximum number of r-adaptive steps on one level of refinement is limited by 21. After every deformation step, the grid quality is enhanced by applying two Laplacian smoothing steps and four grid optimisation steps. In the both rh-AFEM, we set the reduction factor c_r to 1.0, i.e. if the estimated error after an r-adaptive step exceeds the estimated error of the previous step, we proceed with regular refinement of the grid from the current step. For applying the fixed fraction strategy which forms a part of the rh-AFEM 5.2.1, we set $\chi = 0.5$. All rh-adaptive computations start on an equidistant tensor product mesh consisting of 768 elements in 48 macros. Comparing the results obtained by these rh-algorithms with the results previously gained from pure h- or r-adaptivity (cf. figure 5.2.2) we observe that algorithm 5.2.1 does not provide optimal results but suffers from the decay in convergence order like pure h-adaptivity. This comes from the fact that once the deformation is terminated, the element size can only be controlled in a macro-wise manner by macro-wise h-adaptivity. As the macros are refined regularly, the grid at the reentrant corner is locally almost regular (compare figure 5.2.1) which is unsuited for our test problem. Due to this principle drawback, we do not investigate this kind of rh-AFEM any further in this thesis. However, the results for all levels of refinement obtained by algorithm 5.2.1 are by far more accurate than the ones computed with pure h-adaptivity starting on the same mesh. In contrast to this, rh-AFEM 5.2.2 leads to grids which are apparently well-adapted to the underlying problem. This is indicated by the order of convergence with respect to the gradient. The gradient error for the L-domain behaves like $\mathcal{O}(NEL^{0.491})$ in this case, which is very close to the optimal order $\mathcal{O}(NEL^{0.5})$. The order of convergence is computed by taking the gradient errors of the final r-adapted meshes for the different levels of refinement. Moreover, the absolute error equals almost the error observed applying pure r-adaptivity, but does not suffer the loss of convergence order on the finest grid with 12 millions of unknowns, which we noted for pure r-AFEM (compare section 5.1).

The introduction of rh-adaptivity was motivated by the aim to increase to computational speed of pure r-AFEM without sacrifying its superior accuracy.

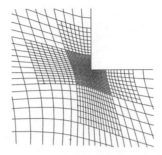

Figure 5.2.1: Excerpt of a grid adapted by *rh*-AFEM 1, test problem 3.6.1

NEL	$||u - u_h||_1$	cycles in r-AFEM	$h_{\min}$
768	$1.63 \cdot 10^{-2}$	21	$3.60 \cdot 10^{-3}$
3,072	$8.23 \cdot 10^{-3}$	8	$9.27 \cdot 10^{-4}$
12,288	$4.16 \cdot 10^{-3}$	7	$2.39 \cdot 10^{-4}$
49,152	$2.10 \cdot 10^{-3}$	6	$6.37 \cdot 10^{-5}$
196,608	$1.06 \cdot 10^{-3}$	5	$1.84 \cdot 10^{-5}$
768,432	$5.38 \cdot 10^{-4}$	4	$5.68 \cdot 10^{-6}$
3,145,728	$2.74 \cdot 10^{-4}$	3	$1.72 \cdot 10^{-6}$
12,582,912	$1.40 \cdot 10^{-4}$	3	$3.95 \cdot 10^{-7}$

Table 5.2.1: Gradient error and minimal mesh width after the final r-adaptive step on a given level of refinement, test problem 3.6.1 with *rh*-AFEM 2

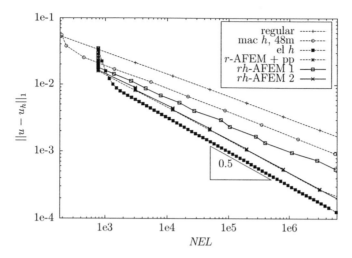

Figure 5.2.2: Gradient error for various adaptivity techniques, test problem 3.6.1

As main bottleneck of r-AFEM we identified that every adaptive iteration takes place on the finest grid in contrast to h-adaptivity. The data in table 5.2.1 shows that in the case of test problem 3.6.1, the number of iterations on the finest grid could be reduced from 17 for the r-AFEM 4.3.1 (compare section 5.1) to only three while maintaining even better accuracy. Thus, we can expect a significant speed-up compared to pure r-AFEM. The overall runtimes presented in figure 5.2.3 for test problem 3.6.1 confirm this. In this figure, we present in the case of rh-AFEM 5.2.2 the runtimes for the final r-deformed grid for the given number of elements, i.e., after the time measurement, the grid is refined globally in the next adaptive iteration. The code is compiled using the Intel Fortran Compiler v. 9.1 with full optimisation and runs on a Opteron 250 server equipped with a 64-bit Linux operating system. For the solve of all Poisson equations in the algorithms either in the test problem itself or inside the deformation algorithm, we employ multigrid which relies on an F-cycle and performs four ADITRIGS-steps [76] for pre- and postsmoothing, respectively. The coarse grid problems are solved by CG-iteration. The parameter settings for both algorithms are the ones described before. We are not able to give timings for the largest problem size considered, because due to memory constraints, these data were gained on another computer with larger main memory, but slower CPU. Thus, these timings do not compare to the ones on the coarser grids. On the finest considered grid with 3 millions of elements, the rh-adaptive algorithm 5.2.2 performs almost 10 times faster than the r-AFEM 4.3.1 with almost identical accuracy. Moreover, it turns out that the runtime grows at lower rate for the rh-AFEM than for the r-AFEM.

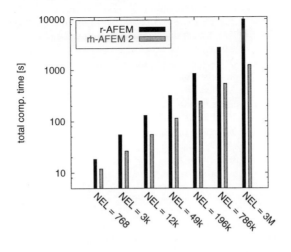

Figure 5.2.3: Total computational time for r-AFEM 4.3.1 with postprocessing (compare section 5.1) and for rh-AFEM 2, test problem 3.6.1

Remark 5.2.3. *To achieve the same accuracy as Rademacher did on an elementwise h-adapted mesh with approx. 6 millions of elements, the rh-AFEM 5.2.2 needs a mesh with roughly 12 millions of unknowns (compare figure 5.2.2). Thus, in the metric NEL vs. accuracy, elementwise h-adaptivity is superior to our rh-AFEM 5.2.2. However, our rh-adaptive calculation with a final mesh consisting of 12 millions of elements takes even on the slower machine mentioned less than one hour. In contrast to this, Rademacher needs for his calculations on a comparable machine approx. 40 hours [64]. Therefore, the rh-AFEM 5.2.2 performs for the test problem calculated by far superior in the metrics CPU time vs. accuracy. However, the suggested speed-up of 40 should not be overinterpreted, as it is influenced by many implementational details like e.g. the choice of the programming language which are not connected to the algorithms themselves but to their practical realisation only.*

5.3 Application of rh-adaptivity to diffusion type problems with heterogenous anisotropic diffusion tensor

For understanding and predicting complex environmental processes, numerical simulation turned out to be an indispensable tool in the last years. In particular, predicting the dispersion of polluted air and the transport of contaminated ground water is of high interest, as without this knowledge, it is difficult up to impossible to develop suitable strategies to react on these problems. The numerical simulation of ground water flow raises the need for reasonable mathematical models of the chemical and fluid flow phenomena occurring in an underground porous medium. These models are to predict the behaviour of *macroscopic* quantities like the averaged pressure p or the averaged volumetric flux of water u. Thus, the microscopic structure of the porous medium is not taken into account in these models. In what follows we are interested in modeling the steady state only and thus neglect any time dependency of the simulated quantities. We additionally assume that there is no air included in the soil such that the problem reduces to modeling single phase flow. Due to the incompressibility of water, the fundamental law of mass preservation states

$$\operatorname{div}(\rho u) = f, \tag{5.1}$$

where f models the presence of sinks or sources of the fluid with density ρ. Darcy's law (see e.g. [11] and the references therein) establishes the relationship between u and pressure gradient ∇p by

$$u = -\frac{K}{\mu}\left(\nabla p - \rho g\right). \tag{5.2}$$

Here, g stands for the acceleration due to gravity, the positive definite 2×2-matrix K describes the absolute permeability of the soil and μ the dynamic viscosity of the water. Due to changes in the underground structure, the coefficients of K may vary rapidly in space. Neglecting g and combining equations (5.1) and (5.2), we end up with

$$-\operatorname{div}\left(\frac{\rho K}{\mu}\nabla p\right) = f. \tag{5.3}$$

Setting $D := \frac{\rho K}{\mu}$, equation (5.3) can be interpreted as generalised Poisson equation:

$$-\operatorname{div}(D \cdot \nabla u) = f, \quad D = \begin{pmatrix} \cos\theta & \sin\theta \\ -\sin\theta & \cos\theta \end{pmatrix}\begin{pmatrix} k_1 & 0 \\ 0 & k_2 \end{pmatrix}\begin{pmatrix} \cos\theta & -\sin\theta \\ \sin\theta & \cos\theta \end{pmatrix}. \tag{5.4}$$

with the corresponding weak formulation

$$(D \cdot \nabla u, \nabla \varphi) = (f, \varphi) \quad \forall \varphi \in H_0^1.$$

For $\theta = 0$, $k_1 = k_2 = 1$, the Poisson equation is recovered. For any tensor D which can be written in the form of (5.4), the generalised Poisson problem is elliptic and thus fulfils the maximum principle. However, for full deformation tensors or general meshes, it is known that many finite difference methods, finite volume methods and even FEM fail to fulfil the discrete maximum principle (DMP) (see

Figure 5.3.1: Subdivision of $[0,1]^2$ in test problem 5.3.1

[59] and the references cited therein). This undesirable property can lead to un-physical behaviour of the solution, e.g. the loss of positivity. Thus, the generalised Poisson equation (5.4) is a challenging problem from the mathematical point of view as well.

In the following, we apply our *rh*-AFEM 5.2.2 to the generalised Poisson equation (5.4). We consider selected test problems defined by Lipnikov et al. [59] in order to investigate the benefit of *rh*-AFEM 5.2.2 with respect to accuracy and positivity of the solution compared to calculations on regular meshes.

Test Problem 5.3.1. *We consider the generalised Possion equation (5.4) on the unit square $[0,1]^2$ with homogenous Dirichlet boundary conditions. As shown in figure 5.3.1, we subdivide the unit square in four congruent squares $\Omega_1, \ldots \Omega_4$. We set $k_1 = 1000, k_2 = 1$ and*

$$\theta = \begin{cases} -\pi/6 & , \quad x \in \Omega_1 \cup \Omega_4 \\ \pi/6 & , \quad x \in \Omega_2 \cup \Omega_3 \end{cases}.$$

The right hand side f is defined as

$$f(x) = \begin{cases} \frac{1}{|\omega|} & , \quad x \in \omega \\ 0 & , \quad x \notin \omega \end{cases}, \quad \omega = [7/18, 11/18]^2.$$

We display the numerical solution of this test problem in figure 5.3.2. As the exact solution is not known, we compute a reference solution on an very fine grid with 16 millions of elements. The true gradient error is then computed by comparison with the recovered gradient of this reference solution. In order to guarantee reliable error computations, we do not consider grids with more than one million of elements for the actual computation on classical tensor product meshes. We stop the *rh*-AFEM 5.2.2 on the final computed grid with 262,144 elements because of the possible existence of very small elements on the deformed grids, where a reliable computation of the gradient error by comparison with the reference solution is hardly possible any more. For the calculations on classical tensor product meshes, we present in table 5.3.1 the gradient error, the efficiency index of its estimation by SPR and the minimal value of the numerical solution

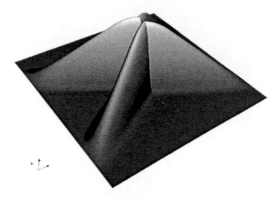

Figure 5.3.2: Numerical solution of test problem 5.3.1

NEL	$||u - u_h||_1$	I_{eff}	$u_{h,\text{min}}$
1,024	$3.63 \cdot 10^{-3}$	0.27	$-2.38 \cdot 10^{-5}$
4,096	$2.26 \cdot 10^{-3}$	0.37	$-2.32 \cdot 10^{-5}$
16,384	$1.19 \cdot 10^{-3}$	0.56	$-1.53 \cdot 10^{-5}$
65,536	$5.83 \cdot 10^{-4}$	0.82	$-6.20 \cdot 10^{-6}$
262,144	$2.83 \cdot 10^{-4}$	1.19	$-8.15 \cdot 10^{-7}$
1,048,576	$1.63 \cdot 10^{-4}$	1.44	$-7.27 \cdot 10^{-11}$

Table 5.3.1: Gradient error, efficiency index and global minimum of the numerical solution for test problem 5.3.1, classical tensor product meshes

$u_{h,\text{min}} := \min_{x \in \Omega} u_h(x)$. Due to the boundary conditions and as the right hand side the exact solution is positive, we measure by $u_{h,\text{min}}$ the violation of the positivity and thus the maximum principle by the numerical method. For sufficiently fine meshes, we observe the expected reduction of the gradient error by a factor of two per refinement which corresponds to an order of $\mathcal{O}(h)$, where h denotes the mesh width. The violation of the positivity is worst on coarse grids and diminishes with successive refinement. The efficiency indices exhibit a stronger variation than in the other test examples considered in this thesis, e.g. the L-domain, where the Poisson equation is involved. However, except on the coarse grids with up to 4,096 elements, the efficiency indices are close to one.

Because the SPR gradient recovery leads to sufficiently accurate estimations of the gradient error for test problem 5.3.1 on tensor product meshes, we rely on this method for the applications of rh-AFEM 5.2.2. For the deformation steps inside the algorithm, we apply the same parameter settings as in the case of test problem 3.6.1, which has been investigated in section 5.2. After every increment of the refinement level, two Laplacian smoothing steps are applied. The maximal

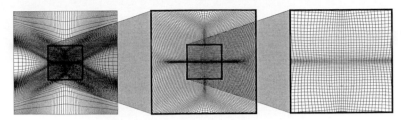

Figure 5.3.3: Grid after 40 adaptive iterations with the rh-AFEM 5.2.2, test problem 5.3.1, $NEL = 65{,}536$

NEL	$\|u - u_h\|_1$	cycles in r-AFEM	I_{eff}	$u_{h,\min}$
1,024	$3.27 \cdot 10^{-3}$	2	0.31	$-2.42 \cdot 10^{-5}$
4,096	$6.34 \cdot 10^{-4}$	10	0.87	$-3.40 \cdot 10^{-6}$
16,384	$2.14 \cdot 10^{-4}$	10	1.13	$-1.14 \cdot 10^{-7}$
65,536	$1.76 \cdot 10^{-4}$	9	0.59	$-3.65 \cdot 10^{-8}$
262,144	$1.47 \cdot 10^{-4}$	2	0.36	$-5.58 \cdot 10^{-9}$

Table 5.3.2: Gradient error, efficiency index and global minimum of the numerical solution for test problem 5.3.1, rh-AFEM 2

number of r-adaptive steps per refinement level is set to 10, the initial grid consists of 1,024 elements. The reduction factor c_r is set to 1.0. One of the resulting grids is displayed in figure 5.3.3. In table 5.3.2, we collect the gradient error, the efficiency indices of the gradient error estimation by SPR and $u_{h,\min}$ after the final r-adaptive step as well as the number of r-adaptive steps on a given level of refinement. The direct comparison of the gradient error and $|u_{h,\min}|$ in figures 5.3.4 and 5.3.5 shows that both quantities benefit from adapting the grids by rh-AFEM 5.2.2. However, on the rather coarse grids with up to 65,536 elements, the improvements are much more pronounced than on the finer grids.

Besides the gradient of the solution, in many application the emphasis is put on the value of quantities derived from the solution itself. As a prototypical problem of this kind, we consider again test problem 5.3.1. In contrast to the previous calculations, we are not interested in the gradient of the solution, but in the point value in $x_e := (0.25, 0.25)$. In these cases, it is decicive not to estimate the gradient error, but the error of the approximate value in the evaluation point x_e. This we achieve employing the DWR-method (for details, we refer to the introduction, section 1.1). With the functional $J_{x_e}(\varphi) := \varphi(x_e)$ and due to the symmetry of D, the dual problem reads

$$(D \cdot \nabla z, \nabla \varphi) = J_{x_e}(\varphi) \quad \forall \varphi \in H_0^1$$

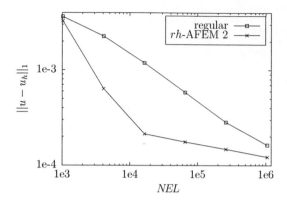

Figure 5.3.4: Gradient error vs. number of elements *NEL* for test problem 5.3.1

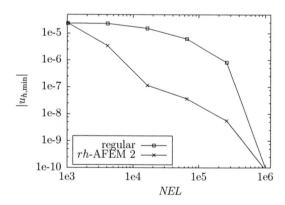

Figure 5.3.5: $|u_{h,\min}|$ vs. number of elements *NEL*, test problem 5.3.1

NEL	$\lvert u - u_h(x_e) \rvert$	I_{eff}	$\lvert u - u_h(x_e) \rvert$	I_{eff}	cycles in r-AFEM
1,024	$1.22 \cdot 10^{-4}$	0.01	$1.16 \cdot 10^{-4}$	0.02	2
4,096	$8.26 \cdot 10^{-5}$	1.46	$2.70 \cdot 10^{-5}$	5.28	2
16,384	$6.18 \cdot 10^{-5}$	1.98	$8.22 \cdot 10^{-6}$	7.03	3
65,536	$2.06 \cdot 10^{-5}$	3.78	$1.97 \cdot 10^{-6}$	1.21	5
262,144	$7.89 \cdot 10^{-6}$	5.01	$2.10 \cdot 10^{-7}$	0.81	5

Table 5.3.3: Point error in $x_e = (0.25, 0.25)$ and efficiency indices for test problem 5.3.1 on classical tensor product meshes (left part) and on grids adapted by rh-AFEM 2 (right part)

and by Galerkin orthogonality, we end up with the error representation

$$J_{x_e}(u - u_h) = (D \cdot \nabla(u - u_h), \nabla(z - z_h)) \,. \tag{5.5}$$

The dual solution z which is depicted in Figure 5.3.6 reveals the influence of the local errors to the error in the evaluation point. In order to obtain a computable error estimation, we replace the unknown terms ∇u and ∇z in formula (5.5) by the reconstructed gradients of u_h and z_h, respectively. For gradient reconstruction, we employ SPR. As the exact solution is unknown, we compute the true error in x_e by comparison with a reference value $\tilde{u}(x_e)$. This value is calculated by extrapolating the corresponding values of the FEM solutions on equidistant tensor product meshes with 1 and 4 millions of elements. This leads to $\tilde{u}(x_e) = 1.199732193 \cdot 10^{-4}$. In order to guarantee reliable error computations, we do not consider grids with more than one million of elements for the actual computation on classical tensor product meshes. When applying rh-adaptivity, we deny meshes with more than 262,144 elements for the same reason. We compute the aforementioned test problem 5.3.1 both on regular tensor product meshes and the sequence of meshes generated by the rh-AFEM 2. We apply the same parameter settings as before, but allow at most 5 subsequent r-adaptive iterations only instead of 10. In table 5.3.3, we present the point error and the efficiency index for the computations on regular tensor product meshes and after the final r-adaptive steps on a given level of refinement. Figure 5.3.7 shows the grid after 12 rh-adaptive steps. Notice the difference to the grid in Figure 5.3.3 which is adapted to the same test problem, but with respect to the gradient error and not the error in a single point. For the unadapted grids with more than 1,024 elements, the point error is increasingly overestimated. It turns out that both the size of the point error and the quality of its estimation benefit considerably from applying rh-AFEM 2. For $NEL = $ 262,144, the point error is more than 30 times smaller on the adapted mesh than on the unadapted one. In contrast to the unadapted case, the quality of the error estimation improves with increasing level of refinement. Thus, we conclude that our rh-AFEM 2 performs well for the computations of derived quantities as well.

We now consider following Lipnikov et al. another test problem which can be regarded as modification of the test problem 5.3.1 we dealt with so far in this

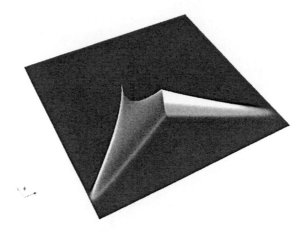

Figure 5.3.6: Numerical dual solution of test problem 5.3.1

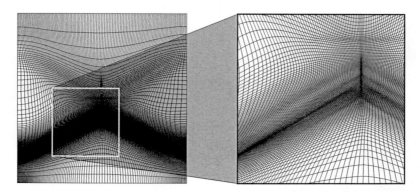

Figure 5.3.7: Grid after 12 adaptive iterations with the rh-AFEM 5.2.2, test problem 5.3.1, point evaluation in $(0.25, 0.25)$, $NEL = 65,536$. For convenience the number of elements has been reduced to 16,384 for display purposes only.

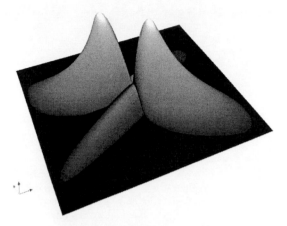

Figure 5.3.8: Numerical solution of test problem 5.3.2

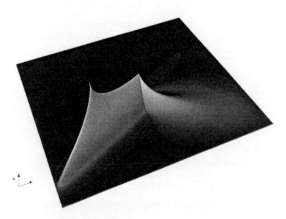

Figure 5.3.9: Numerical dual solution of test problem 5.3.2 with point evaluation in $(0.25, 0.25)$

section.

Test Problem 5.3.2. *We seek to solve the generalised Possion equation (5.4) on the unit square* $[0,1]^2$ *with homogenous Dirichlet boundary conditions. As shown in figure 5.3.1, we subdivide the unit square in four congruent squares* $\Omega_1, \ldots \Omega_4$. *We set*

$$\theta = \left\{ \begin{array}{rl} -\pi/6 & , \quad x \in \Omega_1 \cup \Omega_4 \\ \pi/6 & , \quad x \in \Omega_2 \cup \Omega_3 \end{array} \right. .$$

and

$$k_1 = \left\{ \begin{array}{rl} 1000 & , \quad x \in \Omega_1 \cup \Omega_4 \\ 10 & , \quad x \in \Omega_2 \cup \Omega_3 \end{array} \right. ,$$

$k_2 = 10$. *The right hand side* f *is defined as in test problem 5.3.1.*

We repeat the calculations we performed in the case of test problem 5.3.1 for the modified test problem 5.3.2. We at first compare the results for this test problem computed on regular grids and grids emerging from the rh-AFEM 5.2.2 with respect to the gradient error and the size of the undershoot $u_{h,\min}$. Like for test problem 5.3.1, due to the maximum principle the exact solution is positive in contrast to the numerical solutions. Thus, the size of $u_{h,\min}$ serves as indicator for to what extent the discrete maximum principle is violated in our FEM computations. The exact solution is unknown, thus we utilise the FEM solution computed on an equidistant tensor product mesh with 16 millions of unknowns as reference solution. In the rh-AFEM, we use the very same parameter settings as in the numerical tests before. For regularly refined meshes, we collect in table 5.3.4 the gradient error and the undershoot $u_{h,\min}$. The corresponding efficiency indices refer to the gradient error estimation using SPR. The gradient error estimation is reliable for all levels of refinement, but the efficiency index increases with the finer grids. For sufficiently fine grids with more than 65,536 elements, the size of the undershoot is in the same range as the size of the final resiuduum in the iterative solver. The gradient error on the finest grid considered is even larger than the one on the next coarser grid with 262,144 elements. Such an behaviour is not predicted by FEM theory and raises the suspect that the replacement of the unknown exact solution with the reference solution computed on an extremely fine grid is responsible for this phenomenon.

In table 5.3.5, we present the gradient error, the efficiency index and the size of the undershoot $u_{h,\min}$ as well as the number of r-adaptive steps on a given level of refinement for the calculations of test problem 5.3.2 employing our rh-AFEM 2. Even though the resulting adapted grid features strong local concentrations of the mesh (see figure 5.3.10), the comparison with the corresponding results on regular grids (see table 5.3.4) shows that for this test problem, there is almost no benefit of rh-AFEM with respect to the gradient error. For the grids with 1,024 and 262,144 elements, the gradient error is even larger on the adapted meshes. The size of the undershoot is reduced considerably on the coarse meshes with 1,024 and 4,096 elements, but on the mesh with 65,536 elements, it is more than a factor of thousand larger than on the regular mesh. However, on the adapted grids

NEL	$\|u - u_h\|_1$	I_{eff}	$u_{h,\min}$
1,024	$8.40 \cdot 10^{-3}$	1.27	$-2.80 \cdot 10^{-5}$
4,096	$4.68 \cdot 10^{-3}$	1.59	$-1.60 \cdot 10^{-5}$
16,384	$2.59 \cdot 10^{-3}$	2.11	$-8.68 \cdot 10^{-8}$
65,536	$1.19 \cdot 10^{-3}$	3.13	$-1.17 \cdot 10^{-11}$
262,144	$6.43 \cdot 10^{-4}$	4.08	$-4.98 \cdot 10^{-13}$
1,048,576	$9.58 \cdot 10^{-4}$	1.91	$-2.10 \cdot 10^{-14}$

Table 5.3.4: Gradient error, efficiency index and global minimum of the numerical solution for test problem 5.3.2 on classical tensor product meshes

NEL	$\|u - u_h\|_1$	cycles in r-AFEM	I_{eff}	$u_{h,\min}$
1,024	$9.08 \cdot 10^{-3}$	10	0.77	$-4.06 \cdot 10^{-6}$
4,096	$4.08 \cdot 10^{-3}$	10	0.78	$-6.66 \cdot 10^{-7}$
16,384	$1.91 \cdot 10^{-3}$	10	0.80	$0.00 \cdot 10^{0}$
65,536	$1.14 \cdot 10^{-3}$	9	0.71	$-2.11 \cdot 10^{-8}$
262,144	$9.78 \cdot 10^{-4}$	7	0.43	$0.00 \cdot 10^{0}$

Table 5.3.5: Gradient error, efficiency index and global minimum of the numerical solution for test problem 5.3.2, rh-AFEM 2

consisting 16,384 and 262,144 elements, there is no undershoot at all. Overall, our rh-AFEM does not lead to considerable improvements compared to the results on regular meshes such that it is hard to justify the additional effort of the grid deformation algorithm and the repeated computations on one level of refinement in rh-AFEM for test problem 5.3.2.

We now turn our attention to the point evaluation x_e and apply in the following computations with test problem 5.3.2 the same parameter settings as for the analogous computation with test problem 5.3.1. The reference value $\tilde{u}(x_e) = 3.17524189 \cdot 10^{-4}$ was computed by extrapolation like for test problem 5.3.1 before. The direct comparison between the results on unadapted and adapted meshes in table 5.3.6 reveals that the point error can be significantly reduced by rh-adaptivity. In all cases, the point error estimation is reliable, the efficiency index is (except for the grid with 262,144 elements) close to one. The point error on the adapted mesh with 262,144 elements is highly uncertain and seems to be an artefact from the replacement of the exact value by the reference value. According to the different quantity of interest as before (gradient error and point error), the grids adapted to the minimisation of the point error (figure 5.3.11) differ from the ones adapted to minimise the gradient error (cf. figure 5.3.10).

In the case of the point evaluation, the rh-AFEM 2 produces more accurate results than computations on regular meshes. The gain of accuracy is significant for both test problems. The gradient error however tells a different story. For the first test problem considered, the gradient error is be reduced by rh-adaptivity,

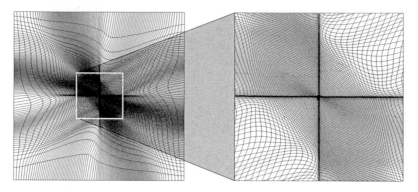

Figure 5.3.10: Grid after 39 adaptive iterations with the rh-AFEM 5.2.2, test problem 5.3.2, $NEL = 65{,}536$. For visualisation purposes only, the number of elements is reduced to 16,384 in this figure.

NEL	$|u - u_h(x_e)|$	I_{eff}	$|u - u_h(x_e)|$	I_{eff}	cycles in r-AFEM
1,024	$4.75 \cdot 10^{-4}$	3.31	$1.39 \cdot 10^{-4}$	1.41	3
4,096	$5.49 \cdot 10^{-5}$	0.59	$4.84 \cdot 10^{-5}$	2.20	2
16,384	$6.21 \cdot 10^{-5}$	0.73	$1.33 \cdot 10^{-5}$	4.28	2
65,536	$3.44 \cdot 10^{-5}$	1.31	$3.99 \cdot 10^{-6}$	3.35	2
262,144	$9.48 \cdot 10^{-6}$	0.91	$7.81 \cdot 10^{-8}$	48.2	3

Table 5.3.6: Point error in $x_e = (0.25, 0.25)$ and efficiency indices for test problem 5.3.2 on classical tensor product meshes (left part) and on grids adapted by rh-AFEM 2 (right part)

but for test problem 5.3.2, there is no gain in accuracy by employing adapted grids. The same holds for the size of the undershoot. This result however does not imply that rh-adaptivity is completely unsuited for this problem. It shows instead that our rh-AFEM 2 combined with error estimation by SPR is not capable to improve the gradient error in this case. Overall, we can conclude that our rh-AFEM 5.2.2 can be successfully applied to generalised Poisson problems with anisotropic diffusion tensor.

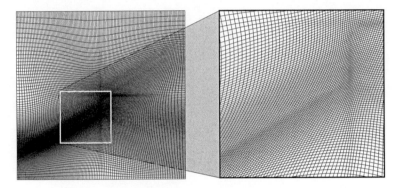

Figure 5.3.11: Grid after 9 adaptive iterations with the rh-AFEM 5.2.2, test problem 5.3.2 and point evaluation, $NEL = 65{,}536$. For visualisation purposes only, the number of elements is reduced to 16,384 in this figure.

Chapter 6

Conclusions and outlook

In this thesis, we developed and analysed a new method for grid deformation. As it preserves the logical structure of the computational grid, it is particularly suitable for grid adaption for high performance computing with the FE package FEAST. One of the reasons for the superior speed of FEAST is the usage of local generalised tensor product meshes. This local structure is preserved by grid deformation. The new method is controlled by a positive *monitor function* which determines the distribution of the element area on the deformed grid. The algorithm itself requires the solution of one global Poisson problem and two IVPs per vertex. These IVPs are fully decoupled. The solution of the IVPs requires searching the grid per ODE time step. For the searching, we proposed and tested raytracing and distance search. It was proven that in a certain sense, the numerical realisation of this method converges with at most first order. We developed extensions of this method with improved accuracy and robustness. Moreover, we presented our multilevel deformation which exploits grid hierarchy and turned to be both convergent of first order and to be of optimal complexity. The grid deformation method was successfully incorporated into an fully r-adaptive algorithm and tested on the L-domain. The sequence of adapted grids allows the approximation of the solution with almost optimal order.

The comparison with macro-wise h-adaptivity and anisotropic refinement illustrated the superiority of our r-adaptive method over the aforementioned ones. Combining h- and r-adaptivity led to the rh-algorithm 5.2.2, which combines both the advantages of h- and r-adaptivity. The application to the L-domain confirmed these properties.

We applied our rh-AFEM additionally to a generalisation of the Poisson equation modeling anisotropic diffusion in porous media. This equation is motivated by the investigation of ground water flow. For the evaluation of the gradient and in particular the evaluation of the solution in a given point, the rh-AFEM 5.2.2 improves the accuracy significantly.

This thesis does by far not mark the endpoint of all research regarding grid deformation techniques, not even the end of research regarding our own method. Although the theoretical foundation of our method is sound and some basic convergence properties understood, it turned out that the convergence theorem 3.1.10 is not optimal, as it predicts the convergence order one only for the quality measure Q_0 whereas we observed the order 1.5 in the corresponding numerical tests. The same tests raised the conjecture that for the sequence of deformed grids, $h_{min} > c h_{max}$ holds even if the preliminaries of lemma 3.1.8 are not fulfilled. A rigorous convergence proof for the multilevel deformation method is still missing. The convergence in the case of the L-domain, where almost none of the assumptions necessary for our convergence theory holds, exhibits the need for a deeper understanding of the convergence phenomena than developed here apparently for a special case of a more general one yet to be formulated.

In some of the tests performed in this thesis, the main drawback of our method became apparent: the lacking possibility to control the shape and/or the position on a single element. Consequently, in some test cases elements occur which feature angles being almost π or almost 0. This undesirable behaviour is connected to the underlying principles of our method, not to any implementational details. Thus, incorporating control over the shape of the elements requires principal modifications and extensions of our deformation method.

On a more practical side, further acceleration of our method can be achieved developing a parallel version of our grid deformation algorithm. The parallelisation of the solver of the Poisson equation inside deformation can be put apart as separate task not restricted to grid deformation and is done already in FEAST. Solving the IVPs in parallel is trivial in theory, as they are not coupled. In fact, it is trivial on a shared-memory architecture, where all data is at hand all time. On a distributed memory system, however, the domain decomposition approach of FEAST makes one node only access the data connected to the subdomain associated to it. In the deformation process, some grid points may leave "their" domain during the IVP solve. This requires communication. The challenge is not to make the deformation method work in parallel, the challenge is to make it work efficiently with minimal amount of communication.

The proper embedding of the new grid deformation method in an r- or rh-adaptive algorithm remains not fully investigated, although we presented first tests and examples in this thesis. These tests and the related research are not exhaustive, but not much more than first trials. In particular, there are no systematic tests yet for different monitor operators M, which define the construction of the monitor function from a given error distribution. On a more fundamental level, there is no proof that our rh-AFEM 5.2.2 converges at all, maybe, this is even wrong.

The successful application of the new rh-adaptive method to the generalised Poisson equation leads to the conjecture that this algorithm can be sucessfully employed to more complex applications like e.g. the Stokes- or Navier-Stokes equations or to problems coming from structural mechanics provided that there are suitable error indicators at hand. We regard the integration of our rh-AFEM in complex FEM simulations as challenging but promising task for the future.

Chapter 7

Appendix

7.1 Proof of equation (2.12)

Lemma 7.1.1. *Formula (2.12) holds.*

Proof. From equation (2.10), we have

$$v(x) = \eta(x, t)\left(t\tilde{f}(x) + (1-t)\tilde{g}(x)\right). \tag{7.1}$$

Applying the chain rule and the div-Operator, we find

$$
\begin{aligned}
\mathrm{div}(v(x)) &= \mathrm{div}\left[\eta(x, t) \cdot \left(t\tilde{f}(x) + (1-t)\tilde{g}(x)\right)\right] \\
&= \left[t\tilde{f}(x) + (1-t)\tilde{g}(x)\right]\mathrm{div}\,\eta(x, t) \\
&\quad + \left(t\nabla\tilde{f}(x) + (1-t)\nabla\tilde{g}(x)\,,\,\eta(x, t)\right).
\end{aligned}
$$

Therefore it follows

$$
\begin{aligned}
\mathrm{div}\,(v(\varphi(x, t))) &= \mathrm{div}\,(\eta(\varphi(x, t), t))\left[t\tilde{f}(\varphi(x, t)) + (1-t)\tilde{g}(\varphi(x, t))\right] \\
&\quad + \left(t\nabla\tilde{f}(\varphi(x, t)) + (1-t)\nabla\tilde{g}(\varphi(x, t))\,,\,\eta(\varphi(x, t))\right). \tag{7.2}
\end{aligned}
$$

Starting from the ODE (2.9), we have by Abel's formula

$$
\begin{aligned}
|J\varphi(x, t)| &= \exp\int_0^t \mathrm{tr}(J\eta(\varphi(x, s), s))ds \\
&= \exp\int_0^t \mathrm{div}\,\eta(\varphi(x, s), s)ds \tag{7.3}
\end{aligned}
$$

and by differentiation of (7.3) we obtain

$$\frac{\partial}{\partial t}|J\varphi(x, t)| = |J\varphi(x, t)|\mathrm{div}\eta(\varphi(x, t), t)). \tag{7.4}$$

147

148

Therefore, we obtain

$$
\begin{aligned}
\frac{\partial}{\partial t} H(x,t) &= \left(\frac{\partial}{\partial t} |J\varphi(x,t)| \right) \cdot \left[t\tilde{f}(\varphi(x,t)) + (1-t)\tilde{g}(\varphi(x,t)) \right] \\
&\quad + |J\varphi(x,t)| \cdot \left[\tilde{f}(\varphi(x,t)) + t \left((\nabla \tilde{f})(\varphi(x,t)), \frac{\partial}{\partial t}\varphi(x,t) \right) \right. \\
&\qquad\qquad \left. -\tilde{g}(\varphi(x,t)) + (1-t) \left((\nabla \tilde{g})(\varphi(x,t)), \frac{\partial}{\partial t}\varphi(x,t) \right) \right] \\
&\stackrel{(7.4)}{=} |J\varphi(x,t)| \cdot \operatorname{div} \eta(\varphi(x,t),t) \cdot \left[t\tilde{f}(\varphi(x,t)) + (1-t)\tilde{g}(\varphi(x,t)) \right] \\
&\quad + |J\varphi(x,t)| \cdot \left[\tilde{f}(\varphi(x,t)) - \tilde{g}(\varphi(x,t)) + \right. \\
&\qquad\qquad \left. \left(t\nabla \tilde{f}(\varphi(x,t)) + (1-t)\nabla \tilde{g}(\varphi(x,t)) \,,\, \frac{\partial}{\partial t}\varphi(x,t) \right) \right] \\
&\stackrel{(2.9)}{=} |J\varphi(x,t)| \cdot \left[\operatorname{div} \eta(\varphi(x,t),t) \left[t\tilde{f}(\varphi(x,t)) + (1-t)\tilde{g}(\varphi(x,t)) \right] \right. \\
&\qquad\qquad + \tilde{f}(\varphi(x,t)) - \tilde{g}(\varphi(x,t)) \\
&\qquad\qquad \left. + \left(t\nabla \tilde{f}(\varphi(x,t)) + (1-t)\nabla \tilde{g}(\varphi(x,t)) \,,\, \eta(\varphi(x,t),t) \right) \right] \\
&\stackrel{(7.2)}{=} |J\varphi(x,t)| \left[\operatorname{div} v(\varphi(x,t)) + \tilde{f}(\varphi(x,t)) - \tilde{g}(\varphi(x,t)) \right] \stackrel{(2.7)}{=} 0
\end{aligned}
$$

$\square$

Bibliography

[1] M. Ainsworth and J. H. Oden. *A posteriori error estimation in finite element analysis*. John Wiley & Sons Ltd., 2000.

[2] M. Altieri, Ch. Becker, S. Kilian, H. Oswald, S. Turek, and J. Wallis. Some basic concepts of FEAST. Preprints SFB 359, Nr. 98-28, Universität Heidelberg, 1998.

[3] I. Babuška. On Besov and Sobolev spaces of fractional order. Technical report, Texas Institute for Comp. and Appl. Mathematics, University of Texas at Austin, 1996.

[4] I. Babuška and M. Dorr. Error estimates for the combined h and p versions of the finite element method. *Numer. Math.*, 37:257–277, 1981.

[5] I. Babuška and W. C. Rheinboldt. A posteriori error estimate for the finite element method. *Int. J. Numer. Methods Eng.*, 12:1597–1615, 1978.

[6] I. Babuška and M. Suri. The p and h-p versions of the finite element method, basic principles and properties. *SIAM Review*, 36(4):578–632, 1994.

[7] W. Bangerth and R. Rannacher. *Adaptive Finite Element Methods for Differential Equations*. Lectures in Mathematics. Birkhäuser, 2003.

[8] R. E. Bank, A. H. Sherman, and A. Weiser. Some refinement algorithms and data structures for regular local mesh refinement. Technical report, Department of Mathematics, University of California at San Diego, La Jolla, California 92093, 1983.

[9] R. E. Bank and A. Weiser. Some a posteriori error estimators for elliptic partial differential equations. *Math. Comput.*, 44:283–301, 1985.

[10] S. Bartels and C. Carstensen. Each averaging technique yields reliable a posteriori error control in FEM on unstructured grids part II: Higher order FEM. *Math. Comp.*, 71:971–994, 2002.

[11] J. Bear. *Dynamics of Fluids in Porous Media*. Dover Publications, New York, 1972.

[12] Ch. Becker. *Strategien und Methoden zur Ausnutzung der High-Performance-Ressourcen moderner Rechnerarchitekturen für Finite-Element-Simulationen und ihre Realisierung in FEAST (Finite Element Analysis & Solution Tools).* PhD thesis, Universität Dortmund, Logos Verlag, Berlin, July 2007.

[13] S. Beuchler. Fast solvers for degenerated problems. Preprints SFB 393, Nr. 03-04, Fakultät für Mathematik, Techn. Universität Chemnitz, February 2003.

[14] H. Blum, J. Harig, S. Müller, and S. Turek. Feat2D - Finite Element Analysis Tools User Manual Release 1.3. Technical report, Universität Heidelberg, January 1992.

[15] H. Blum, A. Schröder, M. Stiemer, Th. Rauscher, H. Kleemann, and A. Rademacher. SOFAR: Small Object oriented Finite element library For Application And Research, 2005. http://www.mathematik.uni-dortmund.de/lsx/research/software/sofar/index.html.

[16] H. Blum and F.-T. Suttmeier. Weighted error estimates for Finite Element solutions of variational inequalities. *Computing*, 65:119–134, 2000.

[17] H. Blum and F.-T. Suttmeier. Adaptive finite elements for contact problems: Case studies in high-speed machining. Ergebnisberichte des Instituts für Angewandte Mathematik, Nr. 201, FB Mathematik, Universität Dortmund, 2001.

[18] P. B. Bochev, G. Liao, and G. C. de la Pena. Analysis and computation of adaptive moving grids by deformation. *Numerical Methods for Partial Differential Equations*, 12:489ff, 1996.

[19] J. U. Brackbill and J. S. Saltzman. Adaptive zoning for singular problems in two dimensions. *J. Comput. Phys.*, 46:342–368, 1982.

[20] D. Braess. *Finite Elements*. Cambridge University Press, 2nd edition, 2001.

[21] J. H. Brandts and M. Křížek. History and future of superconvergence in three-dimensional finite element methods. In *GAKUTO Internat. Ser. Math. Sci. Appl.*, pages 22–33, 2001. Tokyo.

[22] X.-X. Cai, D. Fleitas, B. Jiang, and G. Liao. Adaptive grid generation based on least-squares finite-element method. *Computers and Mathematics with Applications*, 48(7-8):1077–1086, 2004.

[23] W. Cao, W. Huang, and R. D. Russell. A study of monitor functions for two-dimensional adaptive mesh generation. *SIAM Journal on Scientific Computing*, 20(6):1978–1994, 1999.

[24] W. Cao, W. Huang, and R. D. Russell. A moving mesh method based on the geometric conservation law. *SIAM Journal on Scientific Computing*, 24(1):118–142, 2002.

[25] C. Carstensen. All first-order averaging techniques for a posteriori finite element error control on unstructured grids are efficient and reliable. *Math. Comput.*, 73(247):1153–1165, 2003.

[26] C. Carstensen and J. Alberty. Averaging techniques for reliable a posteriori FE-error control in elastoplasticity with hardening. *Comput. Methods Appl. Mech. Eng.*, 192(11-12):1435–1450, 2003.

[27] C. Carstensen and S. Bartels. Each averaging technique yields reliable a posteriori error control in FEM on unstructured grids part I: Low order conforming, nonconforming, and mixed fem. *Math. Comp.*, 71:945–969, 2002.

[28] C. Carstensen, S. Bartels, and S. Jansche. A posteriori error estimates for nonconforming finite element methods. *Numerische Mathematik*, 92:233–256, 2002. DOI 10.1007/s002110100378.

[29] C. Carstensen and S. A. Funken. A posteriori error control in low-order Finite Element discretisations of incompressible stationary flow problems. *Math. Comp.*, 70(236):1353–1381, 2001.

[30] C. Carstensen and S. A. Funken. Averaging technique for FE-a posteriori error control in elasticity. Part I: Conforming FEM. *Comput. Methods Appl. Mech. Engrg.*, 190(18-19):2483–2498, 2001.

[31] C. Carstensen and R. H. W. Hoppe. Convergence analysis of an adaptive nonconforming finite element method. *Numerische Mathematik*, 103:251–266, 2006.

[32] Carstensen, C. and Verfürth, R. Edge residuals dominate a posteriori error estimates for low order finite element methods. *SIAM J. Numer. Anal.*, 36(5):1571–1587, 1999.

[33] R. Courant. Variational methods for the solution of problems of equilibrium and vibrations. *Bull. Amer. Math. Soc.*, 49:1–23, 1943.

[34] E. Creusé, S. Nicaise, and G. Kunert. A posteriori error estimation for the Stokes problem: Anisotropic and isotropic discretizations. *Math. Models Methods Appl. Sci.*, 14(9):1297–1341, 2004.

[35] B. Dacorogna and J. Moser. On a partial differential equation involving Jacobian determinant. *Annales de le Institut Henri Poincaré*, 7:1–26, 1990.

[36] L. Demkowicz and P. Šolin. Goal-oriented *hp*-adaptivity for elliptic problems. *Comput. Methods Appl. Mech. Engrg.*, 193:449–468, 2004.

[37] W. Dörfler. A convergent adaptive algorithm for poisson's equation. *SIAM Journal on Numerical Analysis*, 33(3):1106–1124, 1996.

[38] K. Eriksson, D. Estep, P. Hansbo, and C. Johnson. Introduction to adaptive methods for differential equations. *Acta Numerica*, pages 1–54, 1995.

[39] D. Fleitas, B. Jiang, and G. Liao. Adaptive grid generation based on the least-squares finite element method. Technical report, Dpt. of Math., University of Texas, Arlington, Texas 76019, 2004.

[40] L. Formaggia, S. Micheletti, and S. Perotto. Anisotropic mesh adaptation in computational fluid dynamics: application to the advection-diffusion-reaction and the Stokes problems. *Appl. Numer. Math.*, 51(4):511–533, 2004.

[41] L. Formaggia and S. Perotto. Anisotropic error estimates for elliptic problems. *Numerische Mathematik*, 94:67–92, 2003. DOI 10.1007/s00211-002-0415-z.

[42] L. A. Freitag. On combining laplacian and optimization-based mesh smoothing techniques. Proceeding, Mathematics and Computer Science Division, Argonne National Laboratory, Argonne National Laboratory, Argonne, Illinois 60439, 1997.

[43] L. A. Freitag, M. Jones, and P. Plassmann. An efficient parallel algorithm for mesh smoothing. Proceeding, Mathematics and Computer Science Division Argonne National Laboratory, Mathematics and Computer Science Division Argonne National Laboratory, 1995.

[44] M. B. Giles and N. A. Pierce. Adjoint error correction for integral outputs. In *Error Estimation and Solution Adaptive Discretization in CFD*, NATO/NASA/VKI Lecture Series, pages 39–86, 2001. NASA Ames Research Center, September 10 -14, 2001.

[45] M. B. Giles, N. A. Pierce, and A. Süli. Progress in adjoint error correction for integral functionals. *Computing and Visualization in Science*, 6:113–121, 2004.

[46] D. Göddeke, R. Strzodka, J. Mohd-Yusof, P. McCormick, S. H. M. Buijssen, M. Grajewski, and S. Turek. Exploring weak scalability for FEM calculations on a GPU-enhanced cluster. *Parallel Computing*, 33(10–11):685–699, 2007.

[47] M. Grajewski, M. Köster, S. Kilian, and S. Turek. Numerical analysis and practical aspects of a robust and efficient grid deformation method in the finite element context. Ergebnisberichte des Instituts für Angewandte Mathematik, Nr. 294, FB Mathematik, Universität Dortmund, August 2005.

[48] S. Grosman. Robust local problem error estimation for a singularly perturbed reaction-diffusion problem on anisotropic finite element meshes. Preprints

SFB 393, Nr. 02-07, Fakultät für Mathematik, Techn. Universität Chemnitz, May 2002.

[49] E. Hairer, S. Norsett, and G. Wanner. *Solving Ordinary Differential Equations I*, volume I - Nonstiff Problems. Springer, Berlin, 1987.

[50] V. Heuveline and R. Rannacher. Duality-based adaptivity in the *hp*-finite element method. *J. Numer. Math.*, 2:95–113, 2003.

[51] J. Hoffman and C. Johnson. Adaptive finite element methods for incompressible fluid flow. In *Error Estimation and Solution Adaptive Discretization in CFD*, NATO/NASA/VKI Lecture Series, pages 87–144, 2001. NASA Ames Research Center, September 2001.

[52] J. Hoffman and C. Johnson. Computability and adaptivity in CFD. Technical report, Courant Institute, New York, Courant institute, 2541 Mercer Street, New York, NY-10012, USA, 2004. Encyclopedia of Computational Mechanics.

[53] F. T. Johnson, E. N. Tinoco, and N. J. Yu. Thirty years of development and application of CFD at Boeing Commercial Airplanes, Seattle. *Computers and Fluids*, 34:1115–1151, 2005.

[54] S. Kilian. *Ein verallgemeinertes Gebietszerlegungs-/Mehrgitterkonzept auf Parallelrechnern*. PhD thesis, Universität Dortmund, Logos Verlag, Berlin, September 2001.

[55] G. Kunert. An a posteriori residual error estimator for the finite element method on anisotropic tetrahedral meshes. *Numerische Mathematik*, 86:471–490, 2000.

[56] M. Křížek. Superconvergence phenomena on three-dimensional meshes. *Int. J. Numer. Anal. Model.*, 2:43–56, 2005.

[57] G. Liao and D. Anderson. A new approach to grid generation. *Applicable Analysis*, 44:285–298, 1992.

[58] G. Liao and B. Semper. A moving grid finite-element method using grid deformation. *Numerical Methods for Partial Differential Equations*, 11:603–615, 1995.

[59] K. Lipnikov, M. Shashkov, D. Svyatskiy, and Yu Vassilevski. Monotone finite volume schemes for diffusion equations on unstructured triangular and shape-regular polygonal meshes. *Journal of Computational Physics*, In Press, Accepted Manuscript.

[60] P. Morin, R. H. Nochetto, and G. K. Siebert. Data oscillation and convergence of adaptive FEM. *SIAM Journal on Numerical Analysis*, 38(2):466–488, 2000.

154

[61] P. Neittaanmäki, S. Korotov, and J. Martikainen. *Conjugate Gradient Algorithms and Finite Element Methods*, chapter A Posteriori Error Estimation of 'Quantities of Interest' on 'Quantity-Adapted' Meshes, pages 171–181. Springer, Berlin, 2004.

[62] J. S. Ovall. Asymptotically exact functional error estimators based on supterconvergent gradient recovery. *Numerische Mathematik*, 102:543–558, 2006. BOI: 10.1007/s00211-005-0655-9.

[63] R. Panduranga. Numerical analysis of new grid deformation method in three-dimensional finite element applications. Master thesis, Universität Dortmund, 2006.

[64] R. Rademacher. private communication, 2007.

[65] R. Rannacher. Adaptive Galerkin finite element methods for partial differential equations. *J. Comput. Appl. Math.*, 128:205–233, 2001.

[66] R. Rannacher and R. Becker. A feed-back approach to error control in finite element methods: Basic analysis and examples. *East-West Journal of Numerical Mathematics*, 4:237–264, 1996.

[67] R. Schneiders. Refining Quadrilateral and Hexahedral Element Meshes. In *Proceedings NUMIGRID 96*, 1996.

[68] R. Schneiders, R. Schindler, and F. Weiler. Octree-based generation of hexahedral element meshes. Proceedings 5th International Meshing Roundtable, Pittsburgh, USA, Lehrstuhl für Angewandte Mathematik, insb. Informatik, RWTH Aachen, Ahornstr.55, 52056 Aachen, F.R. Germany, 1996.

[69] A. Schröder. *Fehlerkontrollierte adaptive h- und hp-Finite-Elemente-Methoden für Kontaktprobleme mit Anwendungen in der Fertigungstechnik.* PhD thesis, Universität Dortmund, http://hdl.handle.net/2003/22487, 2005.

[70] F.-T. Suttmeier. Goal-oriented error-estimation and postprocessing for FE-discretisations of variational inequalities. Ergebnisberichte des Instituts für Angewandte Mathematik, Nr. 257, FB Mathematik, Universität Dortmund, May 2004.

[71] F.-T. Suttmeier. Reliable, goal-oriented postprocessing for FE-discretisations. *Numer. Methods Partial Differ. Equations*, 21(2):387–396, 2004.

[72] F.-T. Suttmeier. Adaptive computational methods for variational inequalities based on mixed formulations. *Int. J. Numer. Meth. Engng*, 68:1180–1208, 2006.

[73] S. Turek. Featflow finite element software for the incompressible Navier-Stokes equations: User manual, release 1.1. Technical report, Universität Heidelberg, 1998.

[74] S. Turek, Ch. Becker, and S. Kilian. Hardware-oriented numerics and concepts for PDE software. *Future Generation Computer Systems*, 22(1-2):217–238, 2006.

[75] S. Turek and R. Rannacher. Numerische Methoden für gewöhnliche Differentialgleichungen (Numerische Mathematik II). Vorlesungsskript, 2006.

[76] S. Turek, A. Runge, and Ch. Becker. The FEAST indices - realistic evaluation of modern software components and processor technologies. *Computers and Mathematics with Applications*, 41:1431–1464, 2001.

[77] R. Verfürth. A posteriori error estimates for nonlinear problems. $L^r(0, T; L^p(\Omega))$-error estimates for finite element discretisations of parabolic equations. *Math. Comp.*, 67(224):1335–1360, 1998.

[78] R. Verfürth. Robust a posteriori error estimates for stationary convection-diffusion equations. Technical report, Fakultät für Mathematik, Ruhr-Universität Bochum, D-44780 Bochum, Germany, 2004.

[79] R. Verfürth. Robust a posteriori error estimations for nonstationary convection-diffusion equations. *SIAM Journal on Numerical Analysis*, 43(04):1783–1802, 2005.

[80] L. Wahlbin. Local behavior in finite element methods. In P. G. Ciarlet and J. L. Lions, editors, *Handbook of Numerical Analysis, Vol. II*, pages 420–500. Elsevier, Amsterdam, 1991.

[81] A. Winslow. Numerical solution of the quasi-linear Poisson-equation in a nonuniform triangle mesh. *J. Comput. Phys.*, 1:149–172, 1967.

[82] L. Zhang, T. Strouboulis, and I. Babuška. A posteriori estimators for the FEM: analysis of the robustness of the estimators for the Poisson equation. *Advances in Computational Mathematics*, 15:375–392, 2001.

[83] Z. Zhang. Polynomial preserving recovery for anisotropic and irregular grids. *J. Comput. Math.*, 22(2):331–340, 2004.

[84] Z. Zhang and A. Naga. Validation of the a posteriori error estimator based on polynomial preserving recovery for linear elements. *Int. J. Numer. Methods Eng.*, 61(11):1860–1893, 2004.

[85] Z. Zhang and J. Zhu. Analysis of the superconvergent patch recovery technique and a posteriori error estimator in the finite element method. I. *Comput. Methods Appl. Mech. Eng.*, 123(1-4):173–187, 1995.

[86] Z. Zhang and J.Z. Zhu. Analysis of the superconvergent patch recovery technique and a posteriori error estimator in the finite element method. II. *Comput. Methods Appl. Mech. Eng.*, 163(1-4):159–170, 1998.

[87] O. C. Zienkiewicz and J. Z. Zhu. The superconvergent patch recovery and a posteriori error estimates. part 1: The recovery technique. *Int. J. Numer. Methods Eng.*, 33:1331–1364, 1992.

[88] O. C. Zienkiewicz and J. Z. Zhu. The superconvergent patch recovery and a posteriori error estimates. part 2: Error estimators and adaptivity. *Int. J. Numer. Methods Eng.*, 33:1365–1382, 1992.